Understanding Mathematical Functions and Operations through Graphing

Tong Wu

Printed in the United States of America
ISBN 979-8-9907131-9-2

Understand Mathematical Functions and Operations
through Graphing

ABSTRACT

Unlock the Power of Math Graphs: A Comprehensive Guide to Understanding Functions in 2D and 3D Spaces Dive into the fascinating world of mathematical graphs with this engaging supplement designed to enhance students' comprehension of functions in various coordinate systems. This book delves into the intricacies of Cartesian Coordinates, Polar Coordinates, Cylindrical Coordinates, Spherical Coordinates, and Parametric Coordinates, providing detailed explanations and numerous examples to aid in visualizing and interpreting complex mathematical concepts. From basic functions to advanced graphs, this comprehensive guide equips readers with the necessary tools to navigate and master the realm of math through the lens of graphs. Embark on a transformative journey through the dimensions of mathematics and unravel the mysteries hidden within function graphs.

Wu, Tong
A Supplemental Math Book

Table of Contents

Q and A with Chat-GPT

My Ten Questions to Chat-GPT

1. Q: Write a paragraph on the idea: sets, logic, and reasoning are the foundations to build the mathematics system.
2. Q: Write a paragraph about the history of measuring tools.
3. Q: Write a paragraph about measurement science.
4. Q: Write a paragraph about measuring tools and human beings.
5. Q: Write a paragraph about the relationship between tools and human beings.
6. Q: Help me with a paragraph on the historical development of trigonometric functions.
7. Q: Write a paragraph about the historical transition of the trigonometric functions from the right triangle to the unit circle or analytical functions.
8. Q: Write a few paragraphs about the importance of understanding functions and their graphs
9. Q: Write a few paragraphs about AI, machine learning, and mathematics models

Sets, Logic, and Reasoning

Q: Write a paragraph on the idea: sets, logic, and reasoning are the foundations to build the mathematics system.

A: Sets, logic, and reasoning are the bedrock upon which the entire edifice of human beings constructed mathematical systems. Sets, as defined collections of objects, provide the fundamental building blocks for mathematical structures and relationships. Through the rigorous application of logic, human beings formulate mathematical propositions, analyzed and proven with precision and clarity. This logical framework enables mathematicians to establish the validity of mathematical statements, leading to new theories and theorems. Furthermore, sound reasoning allows mathematicians to infer conclusions from established axioms and principles, fostering a deeper understanding of mathematical concepts and their interconnections. In essence, sets, logic, and reasoning form the indispensable foundation upon which mathematicians build the beauty and complexity of mathematical systems, guiding mathematicians in exploring the infinite realms of mathematical knowledge.

Measuring Tools

Q: Write a paragraph about the history of measuring tools.

A: The history of measuring tools is a fascinating journey that spans thousands of years, starting with the ancient Egyptians, who used tools like cubit rods for measuring length and time. The Greeks later developed more precise measuring instruments, such as the astrolabe for navigation and the water clock for timekeeping. During the Renaissance, Pierre Vernier's invention of the Vernier scale revolutionized precision measurement. The Industrial Revolution mass-produced measuring tools, including micrometers and calipers. In the modern era, technological advancements have developed highly accurate and automated measuring instruments like laser scanners and digital multimeters, shaping how we measure and understand the world.

Measurement,

Q: Write a paragraph about measurement science.

A: Measurement science is a fundamental discipline that underpins a wide range of fields, including physics, engineering, chemistry, and biology. It involves developing, improving, and applying precision and accuracy techniques for quantifying physical quantities such as length, time, mass, and temperature. Through rigorous experimentation, data analysis, and standards development, measurement scientists ensure that measurements are reliable, consistent, and reproducible. The advancements in measurement science drive innovation and technological progress and play a crucial role in shaping our understanding of the natural world and improving our quality of life.

Transaction of Measuring Tools

Q: Write a paragraph about measuring tools and human beings.

A: Measuring tools play a significant role in the relationship between human beings and their surrounding environment. From the simplest ruler to the most sophisticated technology, these tools help humans quantify and understand various aspects of the world around them. Measuring tools facilitate scientific research and technological advancements and impact daily life by influencing decisions and actions. Whether measuring distances, temperatures, weights, or any other parameters, these tools are extensions of human capabilities, enabling us to explore and navigate our surroundings with greater precision and accuracy.

Tools and Human Beings

Q: Write a paragraph about the relationship between tools and human beings.

A: The relationships between tools and human beings are deeply intertwined, dating back to the origins of our species. Tools have played a crucial role in advancing civilization, enabling humans to manipulate their environment, create new technologies, and improve their quality of life. From the earliest stone tools used for hunting and gathering to the sophisticated machinery of the modern world, tools have shaped how humans interact with the world around them. However, this relationship is not one-sided; as humans develop tools, tools also shape humans, influencing their behaviors, thinking processes, and societal structures. This symbiotic relationship between tools and human beings highlights the interconnectedness of innovation, creativity, and progress.

Introduction of Trigonometric Functions

Q: Help me with a paragraph on the historical development of trigonometric functions.

A: Trigonometric functions have a rich historical development dating back to ancient civilizations such as the Babylonians, Egyptians, and Greeks. The Babylonians were among the first to study angles and ratios in triangles, laying the foundation for the development of trigonometry. The Egyptians also used trigonometric concepts in their architectural and surveying practices. However, the Greeks, particularly Hipparchus and Ptolemy, formalized the study of trigonometry by introducing the trigonometric functions we recognize today, such as sine, cosine, and tangent. These functions became essential tools in astronomy, navigation, and mathematics, eventually leading to the sophisticated trigonometric relationships and identities we use extensively in various fields today.

Transition from Geometry to Trigonometric Functions

Q: Write a paragraph about the historical transition of the trigonometric functions from the right triangle to the unit circle or analytical functions.

A: The transition in the historical development of trigonometric functions from the context of right triangles to the unit circle and eventually to analytical functions marked a significant advancement in mathematical understanding. While trigonometry initially focused on relationships within right triangles, the shift toward the unit circle expanded its applications and provided a more general framework. The unit circle allowed for a deeper exploration of trigonometric concepts beyond acute angles, enabling the development of sine and cosine functions as coordinates on the unit circle. This shift paved the way for developing trigonometric identities and extending trigonometric functions to complex numbers. The six trigonometric functions integrate into the category of analytical functions that play a crucial role in modern mathematics and its applications.

Functions and graphs

Q: Write a few paragraphs about the importance of understanding functions and their graphs

A: Understanding functions and their graphs are essential in mathematics as they form the foundational components of many mathematical concepts and real-world applications. Functions describe how one quantity depends on another, providing a framework for analyzing relationships between variables. By studying functions and their graphs, students develop critical thinking skills and improve their problem-solving abilities.

Functions and their graphs are used extensively in various fields, such as physics, engineering, economics, and biology, to model and analyze real-world phenomena. The ability to interpret functions' graphs helps make predictions, identify patterns, and understand systems' behavior. For instance, functions and graphs in physics represent motion, forces, and energy, enabling scientists and engineers to design and optimize systems.

Moreover, understanding functions and their graphs is crucial for calculus, a fundamental branch of mathematics. Calculus relies heavily on the concept of functions and their graphs to study rates of change, areas under curves, and optimization problems. Mastery of functions and their graphs provides a solid foundation for advanced mathematical topics and facilitates the development of critical thinking and analytical skills. In conclusion, a deep understanding of functions and their graphs is vital for academic success and practical applications across various disciplines.

AI

Q: Write a few paragraphs about AI, machine learning, and mathematics models

A: Artificial intelligence (AI) and machine learning have revolutionized various fields, including mathematics, by enabling the development of powerful models for analyzing and processing complex data. One fascinating application of AI in mathematics is the utilization of machine learning algorithms to convert pictures into mathematical models. This innovative approach involves training neural networks to interpret visual data from images and extract relevant mathematical information, creating mathematical representations that can be further analyzed and utilized for various purposes.

By integrating AI and machine learning techniques, researchers and mathematicians can transform visual information captured in pictures into mathematical equations, graphs, or models. This process involves using image recognition algorithms to identify patterns, structures, and relationships in the visual data and then converting them into mathematical representations. These mathematical models can provide insights into the underlying principles governing the observed phenomena in the images, enabling mathematicians to analyze, predict, and manipulate the data more effectively.

By harnessing the power of AI and machine learning to convert pictures into mathematical models, mathematicians can enhance their understanding of complex systems, optimize decision-making processes, and discover new mathematical relationships that may not be apparent through traditional methods. This interdisciplinary approach bridges the gap between visual data and mathematical analysis, opening up new possibilities for innovation and discovery in mathematics. As AI advances, the integration of machine learning for transforming pictures into mathematical models promises to reshape how mathematicians approach problem-solving and data analysis, leading to exciting new insights and applications in mathematics.

Importance of Mathematics

Q: Why is studying mathematics so important?

A: Studying mathematics is paramount for developing essential problem-solving skills applicable across various disciplines and real-life situations. Mathematics is the foundation for multiple fields, such as science, engineering, economics, and computer science. It provides a systematic and logical approach to analyzing and solving complex problems, fostering critical thinking and analytical abilities in individuals. By studying mathematics, one learns how to think logically, make sound decisions based on evidence and data, and approach challenges with a methodical mindset.

Furthermore, mathematics plays a crucial role in our daily lives, from managing personal finances to making informed decisions about health, technology, and the environment. Strong mathematical skills empower individuals to interpret and evaluate information effectively, enabling them to navigate our increasingly data-driven world. Moreover, professions such as finance, statistics, research, and technology heavily rely on mathematical principles for modeling, analysis, and prediction. A solid understanding of mathematics opens up numerous career opportunities and equips individuals with the tools to excel in a rapidly evolving global economy.

In addition, studying mathematics cultivates a growth mindset and resilience in facing complex problems. Solving intricate mathematical concepts encourages perseverance, creativity, and adaptability, which are invaluable in academic and professional settings. Mathematics also promotes curiosity and a deep appreciation for the beauty of patterns, structures, and relationships in the world around us. The study of mathematics enriches the mind and empowers individuals to make informed decisions, solve complex problems, and contribute meaningfully to society's progress and innovation.

Chapter One: Numbers and Basic Operations

 I. Different Sets of Numbers
  1. Natural Numbers, symbol **N**
  2. Whole Numbers, symbol W
  3. Integers, symbol Z
  4. Rational Numbers, symbol Q
  5. Irrational Numbers, symbol Q'
   a. Algebraic Irrational Numbers
   b. Transcendental Irrational Numbers
  6. Real Numbers, R
  7. Complex Numbers, C

 II. Ways of Representing Numbers
  1. Cardinals
  2. Fractions
  3. Decimals
  4. Continued Fractions
   a. Simple Continued Fraction Form
   b. The Finite Continued Fractions and Recursive
    i. Rational Numbers
   c. The Infinite Continued Fractions and Recursive
    i. Quadratic Irrationals
    ii. Transcendental Irrationals
    iii. Construct Irrationals

 III. Relationships Among Numbers
  1. Equivalent
  2. Ordering
  3. Cartesian Products (Basic Operations: +, -, *, /.
  4. Functions as (Roots, Powers)

I. Different Sets of Numbers
1. Natural Numbers, Symbol N

We give counting numbers a new name, natural numbers, which is the definition of natural numbers based on the cardinality of a set. A set is a collection of objects. A set contains one object/element; we say the cardinality of the set is one. The symbol for one is 1. The union of two sets with one object/element has the cardinality of two. The symbol for two is 2. So, we have the concept of natural numbers. The cardinality is a function that maps every finite countable set to the natural numbers. We call the cardinality the n-Function and define it as $n(A): A \xrightarrow{n} N$; where A is a set, and N is the set of natural numbers. Abstractly, we define natural numbers and additions. Natural numbers start with 1 with one operation, plus +, so 1+1=2, 2+1=3, and so on. The operation plus, symbol +, is a binary operation; One, symbol 1, is called the unity. Plus, one, +1, is also called the translation, one of the transformations. Plus (addition) is a cartesian product relation of the natural number sets: $N \times N \xrightarrow{+} N$. We need support from the physical world application when defining abstract mathematical concepts. The support is using the union of two sets.

Example:

1. Let $A = \{a\}$ and $B = \{b\}$, then the union of A and B is $A \cup B = \{a, b\}$. Applying the cardinality function, $n(A) = 1,\ n(B) = 1, and\ n(A \cup B) = 2$, thus we define 1+1=2.
2. Let $A = \{a, c\}$ and $B = \{b\}$, then the union of A and B is $A \cup B = \{a, b, c\}$. Applying the cardinality function, $n(A) = 2,\ n(B) = 1, and\ n(A \cup B) = 3$, thus we define 2+1=3.

$N = N_0 = \{1, 2, 3, ...\}$. The system (N, +) is called a magma. A magma has the closure property. Since the operation plus + has the associativity property, the system is a semigroup, a monoid without the identity 0. A semigroup is a magma where the operation is associative. A monoid is a semigroup with an identity element. Magmas, semigroups, monoids, groups, semirings, rings, and fields are algebraic structures.

Next, we define zero, 0. The cardinality of the empty set $\emptyset$. So n(empty set) = 0.

2. Whole Numbers, Symbol W

The whole number set is the union of {0} and the natural numbers.

$W = N_1 = \{0, 1, 2, 3, ...\}$. The system (W, +) is an unital magma. Since the operation plus symbol + has associativity property, the system is a monoid, a semigroup with identity. The identity is zero, 0.

Let us look at the associativity of +

We defined $1 + 1 = 2$ and $2 + 1 = 3$. How about $1 + 2$? $1 + 2 = 1 + (1 + 1)$. We use the equivalence of two sets to prove that the plus has the associativity. So $1 + (1 + 1) = (1 + 1) + 1 = 2 + 1 = 3$. The answer is $1 + 2 = 3$.

We use the foundations of sets, the union of sets, and the cardinality function to understand the associativity property of addition on whole numbers.

The identity is zero, 0. The zero has the property $0 + a = a = a + 0\ for\ \forall a \in W.\ \forall a \in W$ is read as all whole numbers.

Next, we define the second operation, multiplication, x. The multiplication is a repeated plus/addition.

So, $1 + 1 + 1 + 1 + 1 + 1 + 1 = 6*1=6=1*6$, we need the symbol 6 to tell us six of one adding together. The union of six sets, and each set contains one element.

Let's use the sets, the union of sets, and the cardinality function to illustrate the associativity property of the multiplication. Example $(3+2)+1=3+(2+1)$

We let A={a,b,c}, B={d,e}, and C={f}.

$$(3 + 2) + 1 = (n(A) + n(B)) + n(C) = n(A \cup B) + n(C) = n(A \cup B \cup C) = n(A) + n(B \cup C)$$

$$= n(A) + (n(B) + n(C)) = 3 + (2 + 1).$$ You can prove that the (W,*) has associativity property.

$$1 + 1 + 1 + 1 + 1 + 1 = n\{a\} + n\{b\} + n\{c\} + n\{d\} + n\{e\} + n\{f\} = n\{a, b, c, d, e, f\}$$

Since $n\{a\} + n\{b\} + n\{c\} + n\{d\} + n\{e\} + n\{f\} = 6 * 1$ and $n\{a, b, c, d, e, f\} = 1 * 6$,

thus, $6 * 1 = 1 * 6$. Now, you can prove that the (W,*) has unity, and one is unity.

$W = N_1 = \{0, 1, 2, 3, \dots\}$. The system (W, *) is an unital magma. Since the operation multiplication, * has associativity property, the system is a monoid, a semigroup with identity. The identity is one: 1.

The identity unity, 1, has the property $1 * a = a = a * 1$ for $\forall a \in W$. $\forall a \in W$ is read as all whole numbers.

The multiplication operation is a cartesian product relation of the whole number sets: $W \times W \overset{*}{\to} W$.

Definition of Exponents:

The exponent operation is a cartesian product relation of the whole number sets: $N_0 \times W \overset{Exp}{\to} N_0$ and $\{0\} \times N_0 \overset{Exp}{\to} \{0\}$.

The mathematical expression is n^m, where $n \in N_0$ and $m \in W$. The interpretation for n^m is the product of m factors of n. It means $n^m = n \cdot n \cdot \dots \cdot n$. If $m \in N_0$ then $0^m = 0$.

$$\underbrace{\qquad\qquad}$$
m factors of n

How do we write the cardinality of a set with twelve objects/elements?

Representing Natural Numbers.

We need some symbols for the new numbers. How many symbols do we need? The answer is ten. We call the number system the base ten numeral system. So, we have the ten symbols, also called digits; we already have the two symbols, 0 and 1. We need eight more. So, the other eight symbols are 2, 3, 4, 5, 6, 7, 8, and 9. How do we represent $9*1 + 1$? We define the place values, group them, and write $9*1 + 1 = 10$. The string 10 means that the digit 1 is in the ten place and the zero is in the one place. We write twelve as 12. What is $10*10$? The answer is $10*10 = 100$. The 1 in 100 occupies the place value of one hundred. What is $10*100 = 10*10*10$? The right side of the equation is a repeated multiplication. So, we need to define a new operation called exponent. So, we write

$10*10*10=10\text{^}3$ or $10 \cdot 10 \cdot 10 = 10^3$. Now we can write $8*10\text{^}3 + 5*10\text{^}2 + 3*10 + 1 = 8531$

or $8 \cdot 10^3 + 5 \cdot 10^2 + 3 \cdot 10 + 1 = 8531$; We read the number as eight thousand, five hundred thirty-one. Now, we can use a string of ten digits to represent any whole number. The system is called the positional system. We can combine the set of whole numbers with two operations: addition and multiplication.

The system (W, +, *) is called a semiring. A semiring is a set with two binary operations/compositions. A semiring has a distributive property such that $a * (b + c) = a * b + a * c$, where $a, b, c \in W$.

How do you find the resulting number for $4 + 3 \cdot 5$? We agreed that we should do multiplication first, then addition.

3. Integers, Symbol Z

Is it possible to add a number to 1, and the result is 0? Oh, we need to define the opposite of the unity. We call the opposite of the unity the negative unity; we use the symbol -1. The result of negative unity adding to positive unity is zero, $1 + (-1) = 0$. Also, -1*N = {-1, -2, -3, …}. This set is called negative integers. The union of N, -1*N, and 0 is called the set of integers. $I = N \cup -1 * N \cup \{0\} = \{…, -3, -2, -1, 0, 1, 2, 3, …\}$. We also define the negative unity as the result of the zero subtracting one. That is 0 - 1 = -1. Let n be a natural number; the zero subtract n of unity is $0 – (1 + 1 + 1 + … + 1) = 0 – n*(1) = -n$. We need to extend the cardinality function and the difference of sets to illustrate the negative integers. Let define $n(\emptyset – \{a\}) = -1$ and $n(\emptyset – \{a, b\}) = -2$ and so on.

The system (I, +) is a group. A group is a set of elements with an operation that combines two elements in the set. A group has four properties: 1. Closure: Adding two integers is an integer. 2. Associative: Adding three integers, adding the sum of the first two integers, and then adding the third integer equals adding the first integer and the sum of the last two integers. 3. Identity: Adding the zero and an integer equals adding the integer and the zero equals the integer. 4. Inverses: for every integer, its opposite exists such that adding the integer and its opposite equals the zero. (I, +, *) is a ring. The opposite operation of addition is called subtraction. So, $2 + (-1) = 2 – 1$. What is the result of 2 – 1? The answer is $2 – 1 = 2 + (-1) = 1 + 1 + (-1) = 1 + (1 + (-1))$ (associativity) $= 1 + 0$ (inverses) $= 1$(identity). We extend the exponent operation to $I \times W \xrightarrow{Exp} I$ with the conditions: 1. the zero raised to the zero, 0^0, is undefined; 2. A non-zero number raised to zero is one. Can we extend the exponent operation to $I \times I \xrightarrow{Exp} I$? We need to define a non-zero integer raised to the negative unity. We call the process to find the reciprocal of an integer. The reciprocals of most integers are not integers anymore. Only two integers, one and the negative one, have reciprocals. The reciprocal of one is one. And the reciprocal of the negative one is the negative one.

Now, we have four basic operations on the number systems: addition, subtraction, multiplication, and exponent. All four operations are called algebraic operations.

The operations addition, subtraction, multiplication, and exponent are binary operations.

We need to solve the problem as an example: a group of five people bought five concert tickets. The total cost is $150, and all five tickets are the same. How much each person should pay?

We need a new operation on the numbers, which is division. A new set of numbers begins, and our human beings are rich.

4. Rational Numbers, symbol Q

We define the expression, n^{-1}, as the reciprocals of the natural numbers, $\{n^{-1} \mid n \text{ is a natural number}\}$ Such that the product of a natural number and its reciprocal is the unity, $n^{-1} \cdot n = 1$, $m \cdot n^{-1} = \frac{m}{n}$, and $n^{-m} = \frac{1}{n^m}$. The reciprocal of n also means that unity is divided by n. Now, we introduce and include the division operation in the basic operations. We have five basic operations: addition, subtraction, multiplication, division, and exponent. All five operations are called algebraic operations. Let us define the set $D = \left\{\frac{1}{n} \mid n \text{ is a natural number}\right\}$ and define the rational numbers $Q = I * D = \left\{\frac{m}{n} \mid m, n \text{ are integers}\right\}$. Both algebra systems (Q, +) and (Q-{0}, *) are commutative groups.

The algebra system (Q, +, *) is called a field. Any rational number can be an exponent.

The operations addition, subtraction, multiplication, and division are binary operations. The operation exponent is also a binary operation. There are two numbers. Through a binary operation, the result is a new number. Sometimes, the latest number does not belong to the number set. We need to expand the number set, and then the mathematicians will form a new one.

5. Irrational Numbers, Symbol Q'
a. Algebraic Irrational Numbers

What number squared is 2? The answer is $2^{\frac{1}{2}}$. Any rational number can act as an exponent. $2^{\frac{1}{2}}$ is not a rational number because we cannot express the square root of two as a ratio of two integers. Can we compare $2^{\frac{1}{2}}$ with a rational number? We can compare any two rational numbers. So, we can put them on the number line. The numbers on the right side are larger than on the left. Where can we put $2^{\frac{1}{2}}$ on the number line? We need decimal notation and other notations.

The continued fraction: using sequences

Let $a_1 = 1$ and $a_{n+1} = 1 + \frac{1}{1+a_n}$, then the limit of the nth term is the square root of two. That is $\lim_{n \to \infty} a_n = \sqrt{2}$. See the figure 1

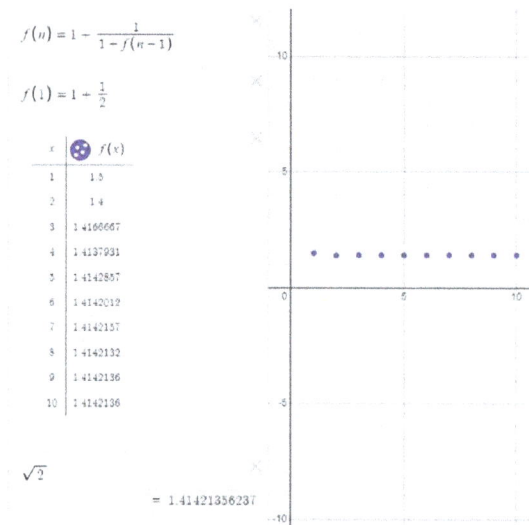

$f(n) = 1 - \frac{1}{1 - f(n-1)}$

$f(1) = 1 + \frac{1}{2}$

x	$f(x)$
1	1.5
2	1.4
3	1.4166667
4	1.4137931
5	1.4142857
6	1.4142012
7	1.4142157
8	1.4142132
9	1.4142136
10	1.4142136

$\sqrt{2}$

$= 1.41421356237$

Figure – 1 using Desmos graphing calculator

$\sqrt{2}$ is a quadratic irrational. It is the principal solution of the quadratic equation: $x^2 - 2 = 0$.

Any rational number can be written as a terminating decimal or repeat decimal.

A proper rational number is in the simplest form $\frac{m}{n}$, the greatest common divisor of m and n is 1.

  i.   If the prime Factorization of n contains only factors 2 or 5, then the rational number can be written as the terminating decimal, such as $\frac{1}{40} = 0.025$.

  ii. If the prime Factorization of n contains a factor other than 2 or 5, then the rational number can be written as the repeating decimal, such as $\frac{1}{7} = 0.[142857]$ with the repetend 142857 and the period of the repetend is 6.

b. Transcendental Irrational Numbers
A transcendental number is a real that is not algebraic.

Examples:

  1. e
  2. π
  3. $\ln 2$

$$e = \lim_{n \to \infty} \left(1 + \frac{1}{n}\right)^n \text{ or } e = \lim_{s \to 0}(1 + s)^{\frac{1}{s}}$$

$$\pi = \lim_{n \to \infty} \sqrt{6 \sum_{i=1}^{n} \frac{1}{i^2}}$$

We can not write any irrational number as a terminating or repeating decimal.

We can use decimals to approximate an irrational number. We need a new notation to equate an irrational number. That is, the continued fraction uses sequences.

6. Real Numbers, Symbol R

The union of rational and irrational is the set of real numbers.

The system (R, +, *) is an algebraic field.

7. Complex Numbers, Symbol C

We may ask what kind of number is $(-1)^{\frac{1}{2}}$? To answer the question, mathematicians expend the Real Number Set to a new number set, Complex Number Set.

A complex number has two parts: one part is a real number, and the second part is an imaginary number.

The form is $a + bi$. Where the i is the $(-1)^{\frac{1}{2}}$. We can use the five operations: addition, subtraction, multiplication, division, and exponent on complex numbers.

The system (C, +, *) is an algebraic field

II. Ways of Representing Numbers

1. Cardinals

We already define the cardinals. We construct various positional systems such as base ten, base two, etc.

2. Fractions

We use the expression of numerator and denominator. Examples of fractions: $\frac{2}{3}, \frac{17}{21}, \frac{101}{120}$.

3. The decimal representation

We can use a decimal to represent any Rational Number. A rational number has either a terminating decimal or a repeating decimal.

Example of the rational number $\frac{103993}{33102}$, it has a terminating decimal.

Its terminating decimal is $3.1415926535898 = 3 \cdot 10^0 + 1 \cdot 10^{-1} + 4 \cdot 10^{-2} + 1 \cdot 10^{-3} + 5 \cdot 10^{-4} + 9 \cdot 10^{-5} + 2 \cdot 10^{-6} + 6 \cdot 10^{-7} + 5 \cdot 10^{-8} + 3 \cdot 10^{-9} + 5 \cdot 10^{-10} + 8 \cdot 10^{-11} + 9 \cdot 10^{-12} + 8 \cdot 10^{-13}$

A terminating decimal is the sum of the product of the digit, face value, and place value.

Let a(i) be a digit of the ten digits. a = [3, 1, 4, 1, 5, 9, 2, 6, 5, 3, 5, 8, 9, 8]

Then $3.1415926535898 = \sum_{i=0}^{13} a[i] \cdot 10^{0-i}$ and $\frac{1}{7} = 142857 \cdot \lim_{n \to \infty} \left(\sum_{i=1}^{n} 10^{0-6i} \right)$.

We use a new infinity symbol ∞, meaning the natural number n gets as large as possible.

We can rewrite the set of natural numbers as N = N$_0$ = {1, 2, 3, ..., n, ...,∞}.

4. Continued Fraction Representation and Recursion

[1] "Every rational number p/q has two closely related expressions as a finite continued fraction. We apply the Euclidean algorithm to (p, q) to determine coefficients a(i). The numerical value of an infinite

continued fraction is irrational. Every irrational number alpha α is the value of a unique infinite regular continued fraction."

A continued fraction can express any real number. The unique canonical finite continued fraction representation can express every rational number. We can express every irrational number in the unique simple infinite continued fraction representation. Every quadratic irrational has a period and repetend.

a. A simple continued fraction form

$$a_0 + \cfrac{1}{a_1 + \cfrac{1}{a_2 + \cfrac{1}{a_3 + \cfrac{1}{\ddots}}}}$$

b. The finite continued fractions and recursive equations

Example 1: a finite continued fraction / rational

Write $\frac{22}{7}, \frac{1}{3}$, and $\frac{31}{25}$ in the canonical representation. And compare it to the repeated decimal representation.

Answer:

a. the canonical representation: $\frac{22}{7} = [3; 7] = 3 + \frac{1}{7}$.

$$A_1(n) = 3 + \frac{1}{A1(n-1)} \; with \; A_1(1) = 7$$

$$A_1(2) = \frac{22}{7}$$

The repeated decimal representation: $\frac{22}{7} = 3.[142857]$. Its period is six and the repetend is 142857.

b. the canonical representation for $\frac{1}{3} = [0; 1] = \frac{1}{3}$.

$$A_1(n) = 0 + \frac{1}{A1(n-1)} \; with \; A_1(1) = 3$$

$$A_1(2) = \frac{1}{3}$$

The repeated decimal representation: $\frac{1}{3} = 0.[3]$. Its period is one and the repetend is 3.

c. the canonical representation: $\frac{31}{25} = [1; 4, 5, 1] = 1 + \cfrac{1}{4 + \cfrac{1}{5 + \frac{1}{1}}}$.

Let the list $L = [a_n, a_{n-1}, a_{n-2}, \dots, a_1, a_0]$ so $L_2 = [1, 5, 4, 1]$ (read the canonical representation from right to left)

$$A_3(n) = L_2[n] + \frac{1}{A_3(n-1)} \; with \; A_3(1) = 1$$

$$A_3(1) = a_n = 1$$

$$A_3(2) = a_{n-1} + \frac{1}{a_n} = 5 + \frac{1}{1} = 6$$

$$A_3(3) = a_{n-2} + \frac{1}{a_{n-1}} = 4 + \frac{1}{6} = \frac{25}{6}$$

$$A_3(4) = a_{n-3} + \frac{1}{a_{n-2}} = 1 + \frac{6}{25} = \frac{31}{25}$$

The number $\frac{31}{25}$ has the terminating decimal representation: $\frac{31}{25} = 1.25$.

14

c. The Infinite Continued Fractions and Recursive Equations
i.   Quadratic irrationals are solutions to a quadratic equation.

Example 2: infinite continued fractions / quadratic irrational

Quadratic irrationals are solutions of a quadratic equation with rational coefficients.

Write $\sqrt{2}$, $\sqrt{19}$, and $\frac{1+\sqrt{5}}{2}$ in the canonical representation. And compare them with the decimal representation.

a. the canonical representation: $\sqrt{2} = [1; [2]]$. Its period is one and the repetend is 2.

The recursive equation is

$$A_1(n) = 1 + \frac{1}{A1(n-1)} \ with \ A_1(1) = 2$$

$$\sqrt{2} = \lim_{n \to \infty} A_1(n)$$

$\sqrt{2}$ is irrational, and its decimal representation is neither terminal nor repeating.

$\sqrt{2} = 1.41421356237 \dots$. It does not have a period nor a repetend.

The expanded canonical representation is $1 + \cfrac{1}{2 + \cfrac{1}{2 + \cfrac{1}{2 + \cfrac{1}{\ddots}}}}$

b. the canonical representation for $\sqrt{19} = [4; [213128]]$. Its period is six, and the repetend is 213128.

Let the list $L = [a_1, a_2, \dots, a_n]$ so $L_2 = [2, 1, 3, 1, 2, 8]$, the list contains all the digits in the repetend and with the order of the digits reading from left to right. $p = 6$ (The period) and $a_0 = 4$ (the whole part).

$$A_2(n) = L_2[6 - mod(n-1,6)] + \frac{1}{A_2(n-1)}$$

$$A_2(1) = L_2[p]$$

$$\sqrt{19} = a_0 + \frac{1}{\lim\limits_{n \to \infty} A_2(pn)}$$

$\sqrt{19}$ is irrational, and its decimal representation is nonterminal and non-repeat.

$\sqrt{19} = 4.35889894354 \dots$. It does not have a period nor a repetend.

The expanded canonical representation is $4 + \cfrac{1}{2 + \cfrac{1}{1 + \cfrac{1}{3 + \cfrac{1}{1 + \cfrac{1}{2 + \cfrac{1}{8 + \frac{1}{\ddots}}}}}}}$

c. the canonical representation: $\frac{1+\sqrt{5}}{2} = [1; [1]] = 1 + \cfrac{1}{1 + \cfrac{1}{1 + \frac{1}{\ddots}}}$.

Its period is one and the repetend is 1.

Let the list $L = [a_1, a_2, \ldots, a_n]$ so $L_3 = [1]$. The list contains all the digits in the repetend, with the order of the digits reading from left to right. $p = 1$ (The period) and $a_0 = 1$ (the whole part).

$$A_3(n) = L_3[1 - mod(n-1, 1)] + \frac{1}{A_3(n-1)}$$

$$A_3(1) = L_3[p]$$

$$\frac{1+\sqrt{5}}{2} = a_0 + \frac{1}{\lim\limits_{n \to \infty} A_3(pn)}$$

$\frac{1+\sqrt{5}}{2}$ is irrational, and its decimal representation is neither terminal nor repeating.

$\frac{1+\sqrt{5}}{2} = 1.618033988749895\ldots$. It does not have a period nor a repetend.

$\frac{1+\sqrt{5}}{2}$ is called the golden ratio. And it is the positive solution of $x^2 - x - 1 = 0$

$\frac{1+\sqrt{5}}{2}$ is ratio of $\frac{y}{x}$ such that $y > x > 0$ and $\frac{x+y}{y} = \frac{y}{x}$.

We use the Greek capital letter $\Phi = \frac{1+\sqrt{5}}{2}$.

The graph of the golden ratio rectangle:

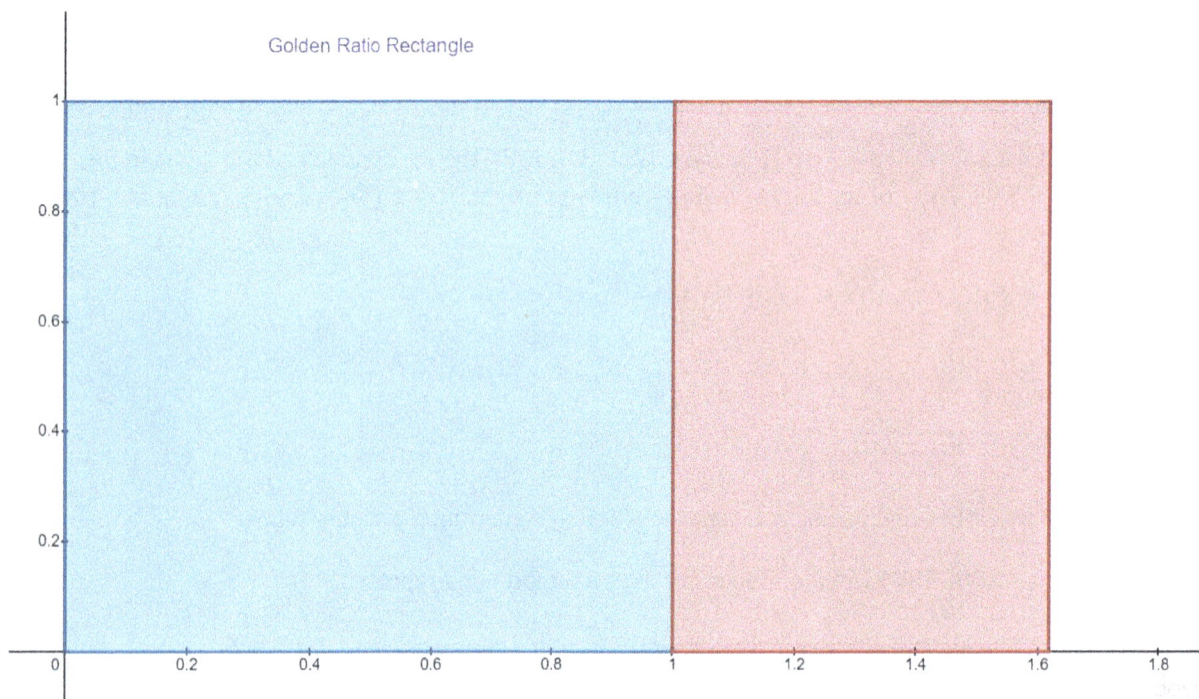

Golden Ratio Rectangle

Summary

The recursive equation for rational numbers

Let L be a list such that $L = [a_1, a_2, \dots a_N]$, and the sequence $A(n) = L[N - n + 1] + \frac{1}{A(n-1)}$ with the first term being $A(1) = L[N]$

Let $\frac{p}{q}$ be a proper fraction, then $\frac{p}{q} = A(N)$.

The recursive equation for quadratic irrational numbers

Let L be a list such that $L = [a_1, a_2, \dots a_p]$ containing all digits of the repetend and the whole part is a_0:

the sequence

$$A(n) = L[p - mod(n - 1, p)] + \frac{1}{A_2(n-1)}$$

$$A(1) = L[p]$$

The number value $= a_0 + \dfrac{1}{\lim\limits_{n\to\infty} A(pn)}$

ii.   The Infinite Continued Fractions and Recursive Equations: Transcendental Irrationals

Example 3: Write e, π, and γ (Euler constant) in the canonical representation. And compare them with the decimal representation.

a. The canonical representation for $e = [2; [1\ 2n\ 1]]$. Its period is three, and the repetend has a pattern but is not fixed: $1\ 2n\ 1$.

Define a piecewise function:

$$f_{31} = \{mod(x, 3) = 1: 1, mod(x, 3) = 2: 2(N - floor\left(\tfrac{x}{3}\right), mod(x, 3) = 0,1)$$

N is a Natural number. For round to 7 digits, $N \geq 5$.

It is equivalent to $L_{31} = [1, 2n, 1]$, the list contains all the digits in the repetend and with the order of the digits reading from left to right. $p = 3$ (The period) and $a_0 = 2$ (the whole part).

$$A_{31}(n) = f_{31}(n - 1) + \frac{1}{A_{31}(n-1)} \ \{n \leq Np + 1\}$$

$$A_{31}(1) = 1$$

$e = a_0 + \dfrac{1}{\lim\limits_{N\to\infty} A_2(Np)}$

e is a transcendental irrational, and its decimal representation is nonterminal and non-repeat.

$e = 2.718281828459\dots$. It does not have a period nor a repetend.

The expanded canonical representation is $2 + \cfrac{1}{1+\cfrac{1}{2+\cfrac{1}{1+\cfrac{1}{1+\cfrac{1}{4+\cfrac{1}{1+\cdots}}}}}}$

b. The canonical representation for $\pi \approx$
[3; [7,15,1,292,1,1,1,2,1,3,1,14, 3, 3,23,1,1,7,4,35,1,1,1,2,3,3,3,3,1,1,14,6,4,5,1,

7,1,5,1,1,3,18,2,1,2,4,2,96,2,3,2,1,1,6,1,6,2,5,64,1,2,3,1,17,5,1,12,3,2,1,

1,1,1,2,2,1,4,1,1,2,2,22,1,2,1,6,1,16,1,2,3,2,4,2,5,2,3,1,1,3,2,1,7,6,4,

4,3,1,61, ...]]. No pattern has been found [1]. Its period is undefined.

Depending on the rounding, Desmos graphing calculator rounds to 12 digits. The above pi canonical representation list differs from A001203's simple continued fraction expansion of Pi by using the desmos graphing calculator to calculate

A001203 $\pi =$
[3; [7, 15, 1, 292, 1, 1, 1, 2, 1, 3, 1, 14, 2, 1, 1, 2, 2, 2, 2, 1, 84, 2, 1, 1, 15, 3, 13, 1, 4, 2, 6, 6, 99, 1,

2, 2, 6, 3, 5, 1, 1, 6, 8, 1, 7, 1, 2, 3, 7, 1, 2, 1, 1, 12, 1, 1, 1, 3, 1, 1, 8, 1, 1, 2, 1, 6, 1, 1, 5,

2, 2, 3, 1, 2, 4, 4, 16, 1, 161, 45, 1, 22, 1, 2, 2, 1, 4, 1, 2, 24, 1, 2, 1, 3, 1, 2, 1, ...]][2].

Let's use the finite simple continued fraction 98 recursive expansion to approximate Pi.

Let the list $L = [a_1, a_2, \dots, a_n]$ so $L_4 =$
[7, 15, 1, 292, 1, 1, 1, 2, 1, 3, 1, 14, 2, 1, 1, 2, 2, 2, 2, 1, 84, 2, 1, 1, 15, 3, 13, 1, 4, 2, 6, 6, 99, 1,

2, 2, 6, 3, 5, 1, 1, 6, 8, 1, 7, 1, 2, 3, 7, 1, 2, 1, 1, 12, 1, 1, 1, 3, 1, 1, 8, 1, 1, 2, 1, 6, 1, 1, 5,

2, 2, 3, 1, 2, 4, 4, 16, 1, 161, 45, 1, 22, 1, 2, 2, 1, 4, 1, 2, 24, 1, 2, 1, 3, 1, 2, 1].

The length is 97 and $a_0 = 3$ (the whole part).

$A_4(n) = L_4[97 - (n-1)] + \cfrac{1}{A_4(n-1)}$

$A_4(1) = L_4[97]$

$\pi \approx 1 + \cfrac{1}{A4(97)}$

π is a transcendental irrational, and its decimal representation is nonterminal and non-repeat.

$\pi = 3.141592653589793 \dots$. It does not have a period nor a repetend.

The expanded canonical representation is $3 + \cfrac{1}{7+\cfrac{1}{292+\cfrac{1}{1+\cfrac{1}{1+\cfrac{1}{1+\cfrac{1}{2+\cdots}}}}}}$

b. The canonical representation for A002852 $\gamma \approx$
$[0; [1,1,2,1,2,1,4,3,13,5,1,1,8,1,2,4,1,1,40,1,11,3,7,1,7,1,1,5,1,49,4,1,65,1,4,$

$$7,11,1,399,2,1,3,2,1,2,1,5,3,2,1,10,1,1,1,1,2,1,1,3,1,4,1,1,2,5,1,3,6,2,1,$$

$2,1,1,1,2,1,3,16,8,1,1,2,16,6,1,2,2,1,7,2,1,1,1,3,1,2,1,2, \dots]$ [3]. No pattern has been found [1]. Its period is unknown. It is an open question: "Is γ a rational or irrational?"

Let's approximate gamma by using the finite simple continued fraction 98 recursive expansion.

Let the list $L = [a_1, a_2, \dots, a_n]$ so $L_5 =$
$[1,1,2,1,2,1,4,3,13,5,1,1,8,1,2,4,1,1,40,1,11,3,7,1,7,1,1,5,1,49,4,1,65,1,4,$

$$7,11,1,399,2,1,3,2,1,2,1,5,3,2,1,10,1,1,1,1,2,1,1,3,1,4,1,1,2,5,1,3,6,2,1,$$

$$2,1,1,1,2,1,3,16,8,1,1,2,16,6,1,2,2,1,7,2,1,1,1,3,1,2,1,2]$$

The length is 98 and $a_0 = 0$ (the whole part).

$$A_5(n) = L_5[98 - (n-1)] + \frac{1}{A_4(n-1)}$$

$$A_5(1) = L_4[98]$$

$$\gamma \approx 0 + \frac{1}{A4(98)}$$

γ is not known as irrational or rational.

$$\gamma = \lim_{n\to\infty} \left(-\ln n + \sum_{i=1}^{n} \frac{1}{n}\right) = 0.577215664901532 \dots . \text{ We do not know whether it has a}$$
period or not.

$$\text{The expanded canonical representation is } 0 + \cfrac{1}{1+\cfrac{1}{1+\cfrac{1}{2+\cfrac{1}{1+\cfrac{1}{2+\cfrac{1}{1+\cfrac{1}{\ddots}}}}}}$$

iii.   Construct irrationals

a.

Let $g(x) = 2x - 1$ and the sequence $A(n) = g(N - (n-1)) + \frac{1}{A(n-1)}$ with the first term is

$A(1) = 2N - 1,$

$A(N)$ is a finite continued fraction and is a rational number.

19

$$A(N) = 1 + \cfrac{1}{3 + \cfrac{1}{5 + \cfrac{1}{\ddots \cfrac{1}{2N-3 + \cfrac{1}{2N-1}}}}}$$

But $\lim_{N \to \infty} A(N)$ is an infinite continued fraction. And it is an irrational number. (See figures 4 & 5)

The abbreviated notation for the number A(N) = [1; 3, 5, ...2N-1] and

$$\lim_{N \to \infty} A(N) = [1; 3, 5, 7, ...].$$

b.

Let $g_2(x) = x$ and the sequence $A(n) = g(N - (n-1)) + \frac{1}{A(n-1)}$ with the first term is

$A(1) = N$,

A(N) is a finite continued fraction. And it is a rational number.

$$A(N) = 1 + \cfrac{1}{2 + \cfrac{1}{3 + \cfrac{1}{\ddots \cfrac{1}{N-1 + \cfrac{1}{N}}}}}$$

But $\lim_{N \to \infty} A(N)$ is an infinite continued fraction. And it is an irrational number. (See figures 4 & 5)

The abbreviated notation for the number A(N) = [1; 2, 3, ...N] and

$$\lim_{N \to \infty} A(N) = [1; 2, 3, 4, ...].$$

For any natural number N, A(N) is rational. And $\lim_{N \to \infty} A(N)$ is irrational, and the period is infinite.

A finite continued fraction (or terminated continued fraction) can represent any rational number.

An infinite continued fraction can represent any irrational number.

III.   Relationships Among Numbers
   1. Equivalent
   2. Ordering
   3. Cartesian Products (Basic Operations: +, -, *, /.
   4. Functions as (Roots, Powers)

   1. Equivalent

Given any two numbers, we should be able to tell whether the two numbers are equivalent or not. Two numbers are equivalent if and only if the two numbers are equal. Because equal is a relation of a set of numbers. And equal is reflexive on the set of numbers, which means any number equals itself; equal is symmetric, which means two numbers a and b a=b if and only if b=a; equal is transitive, which means three numbers a, b, and c, if a = b and b = c, then a = c.

   2. Ordering

If two numbers are not equivalent, we compare the two; one number is smaller than the other.

   3. Cartesian Products (Basic Operations: +, -, *, /.

The four basic operations, addition, subtraction, multiplication, and division, are cartesian products.

The whole number sets: $W \times W \xrightarrow{+} W$, $W \times W \xrightarrow{*} W$

The integer sets: $Z \times Z \xrightarrow{+} Z$, $Z \times Z \xrightarrow{-} Z$, $Z \times Z \xrightarrow{*} Z$

The rational sets: $Q \times Q \xrightarrow{+} Q$, $Q \times Q \xrightarrow{-} Q$, $Q \times Q \xrightarrow{*} Q$, $Q \times Q \xrightarrow{\div} Q$

The real number sets: $R \times R \xrightarrow{+} R$, $R \times R \xrightarrow{-} R$, $R \times R \xrightarrow{*} R$, $R \times R \xrightarrow{\div} R$

The complex number sets: $C \times C \xrightarrow{+} C$, $C \times C \xrightarrow{-} C$, $C \times C \xrightarrow{*} C$, $C \times C \xrightarrow{\div} C$

Definition of the Cartesian Product

Let A and B be sets. The set of all ordered pairs having the first coordinate in A and a second coordinate in B is called the Cartesian product of A and B and is written $A \times B$. Thus

$$A \times B = \{(a, b): a \in A \text{ and } b \in B\}$$

4. Functions as (Roots, Powers)

Definition: a function (or mapping) from A to B is a relation f from A to B such that

1. the domain of f is A.

2. if $(x, y) \in f$ and $(x, z) \in f$, then $y = z$.

We write $f: A \to B$, and we read "f is a function from A to B," or "f maps A to B."

The set B is called the codomain of f. In the case where $A = B$, we say f is a function on A.

The Roots and Powers are functions from the number set to a number set.

Chapter Two: Functions

I.   Discrete Functions
II.   Continuous Functions
III.   Algebraic Functions
IV.   Transcendental Functions

Discrete Functions

The domain is a subset of the set of integers. We call the set of integers the support, environmental, universal, or ambient sets of a discrete function.

Examples:

1. Binomial Distribution
2. Geometric Distribution
3. Poisson Distribution
4. Hypergeometric Distribution
5. Sequences

Continuous Functions

The domain is a subset of the set of real numbers. We call the set of real numbers the support, environmental, universal, or ambient sets of a continuous function.

Examples:

1. Polynomial Functions
2. Normal Distribution
3. Student's t-Distribution
4. Chi-Square Distribution
5. F Distribution
6. Gamma Function
7. Gamma Distribution

Algebraic Functions

Examples:

1. Polynomial Functions
2. Rational Functions
3. Cauchy Distribution

Transcendental Functions

Examples:

1. Trigonometric Functions
2. Exponential Functions
3. Logarithmic Functions

I. Discrete Functions
  1. Binomial Distribution
  2. Geometric Distribution
  3. Poisson Distribution
  4. Hypergeometric Distribution
  5. Sequences

I. Discrete Functions

1. Binomial Distribution

$$B(x,n,p) = C_x^n \cdot p^x \cdot (1-p)^{n-x}, x = \{0,1,\ldots,n\}$$

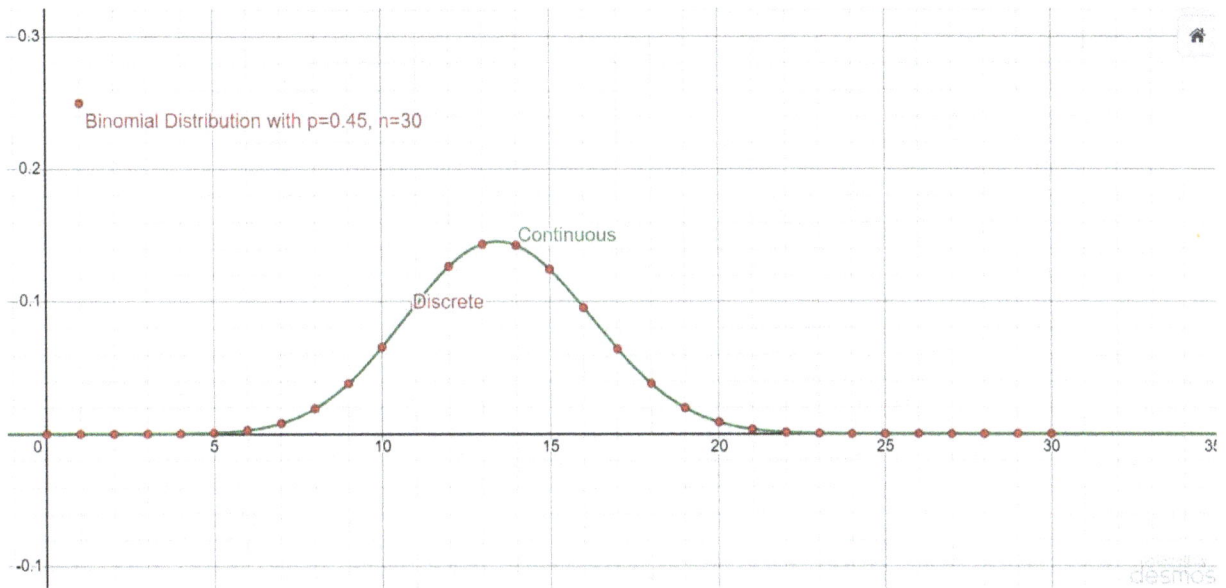

Bernoulli Distribution (special case of binomial distribution: n=1

$$B_1(x,p) = C_x^1 \cdot p^x \cdot (1-p)^{n-x}, x = \{0,1\}$$

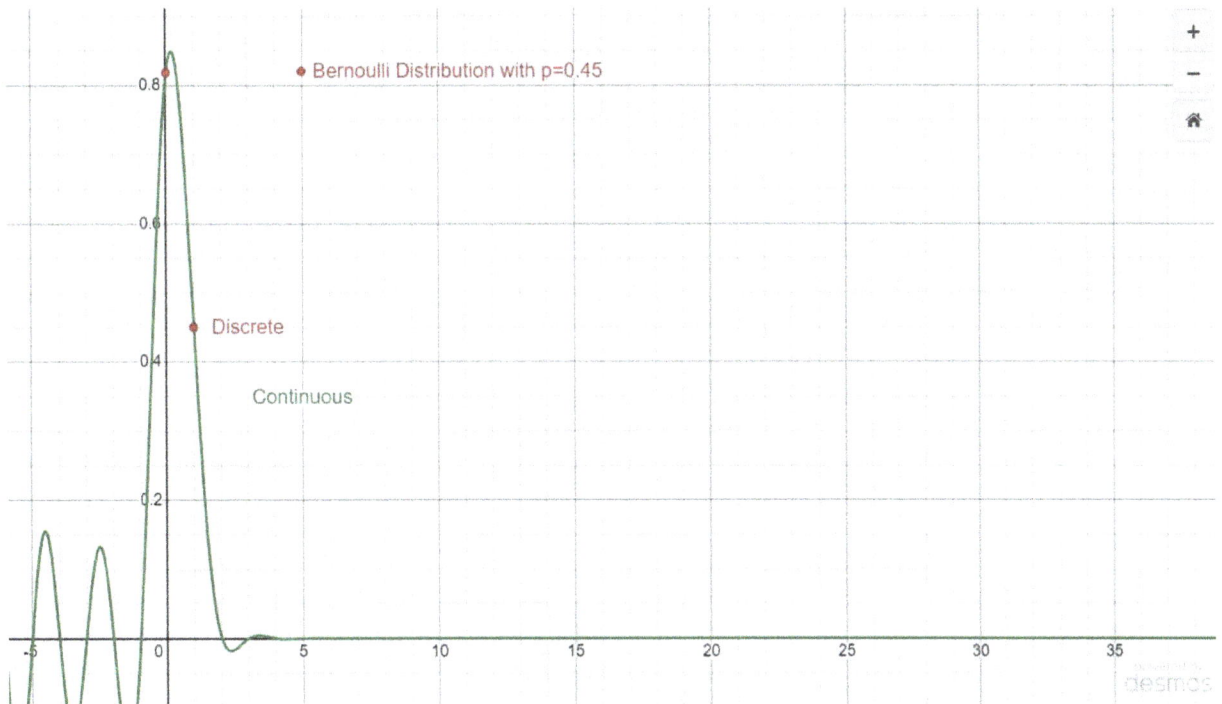

2. Geometric Distribution

$$G(x, p) = p(1-p)^{x-1}, \qquad x = \{1, 2, \dots\}$$

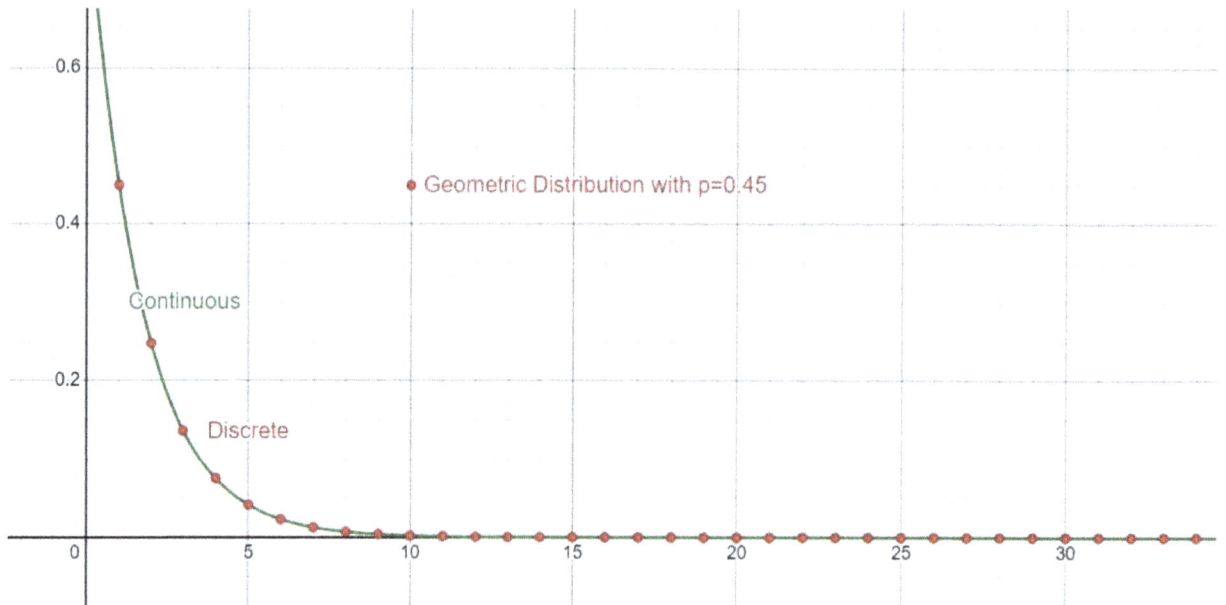

3. Poisson Distribution

$$P(x, \lambda) = \frac{e^{-\lambda} \cdot \lambda^x}{x!}, x = \{0, 1, 2, \dots\}$$

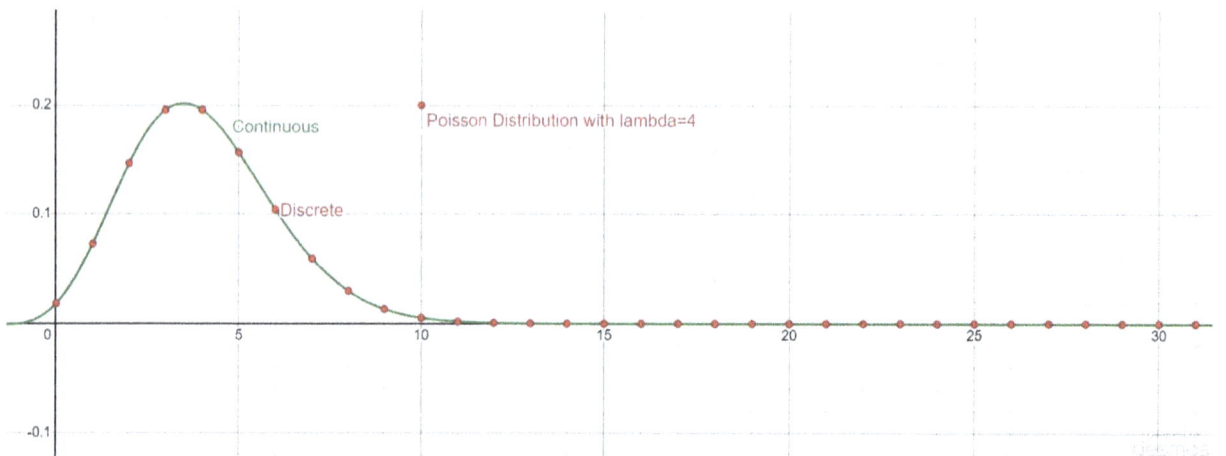

4. Hypergeometric Distribution

$$H(x, n, N, k) = \frac{C_x^k \cdot C_{n-x}^{N-k}}{C_n^N}, x = \{0, 1, \dots, n\}$$

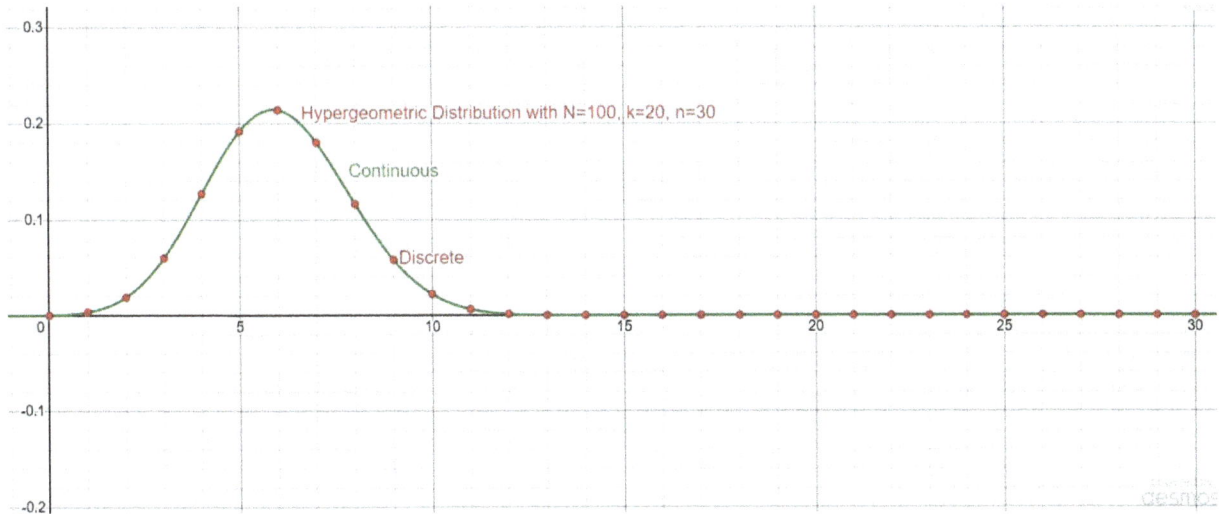

5. Sequences
   a. Arithmetic Sequences

$$a_n = a_1 + (n-1)d$$

$$a_x = A(x, a_1, d) = a_1 + (x-1)d, x = \{1, \dots, n, \dots\}$$

Example:

$$a_n = A(n, -4, 3) = -4 + (n-1) \cdot 3 = 3n - 1, n = \{1, 2, 3, \dots\}$$

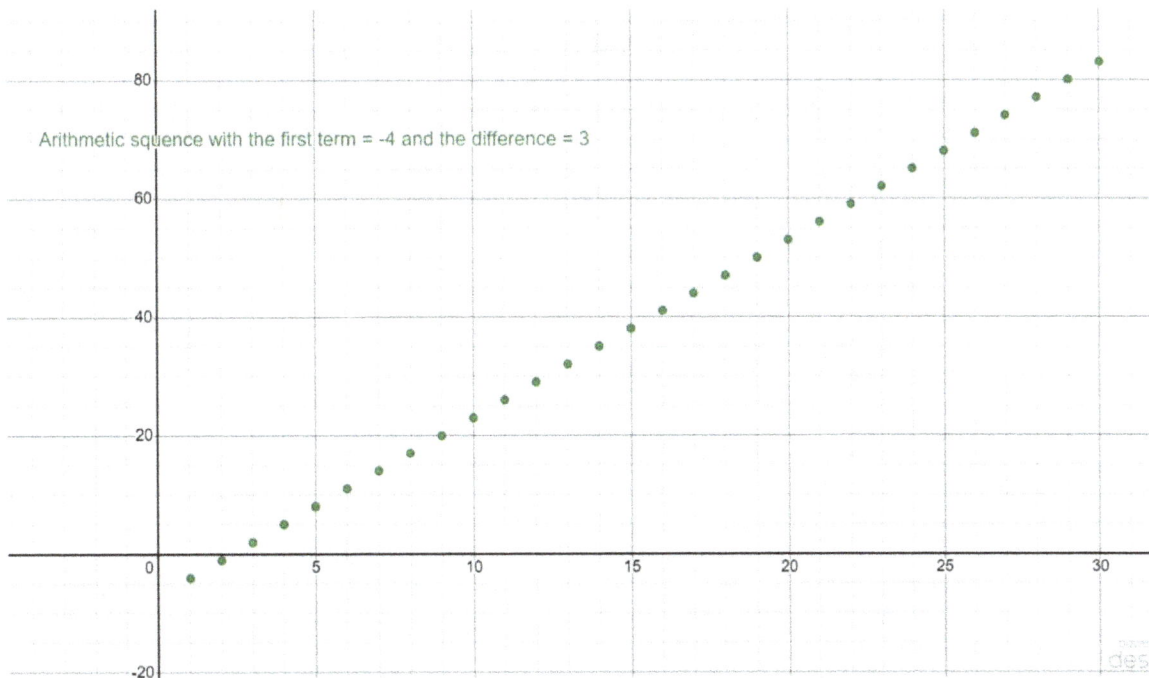

b. Geometric Sequences

$$a_n = a_1 \cdot r^{n-1}$$

$$a_x = G(x, a_1, r) = a_1 \cdot r^{x-1}, x = \{1, \dots, n, \dots\}$$

Example b1:

$$a_n = G(n, 5, 2) = 5 \cdot 2^{n-1}, n = \{1, 2, 3, \dots\}$$

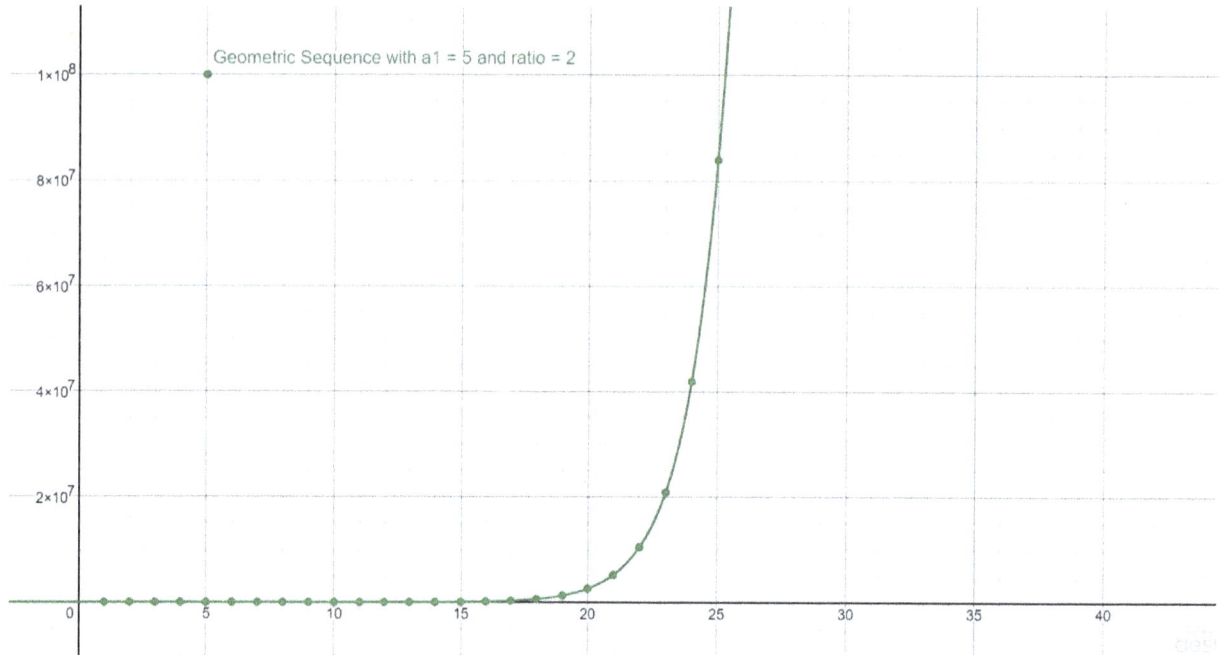

Geometric Sequence with a1 = 5 and ratio = 2

Taking logarithm: $\ln a_n = \ln(5 \cdot 2^{n-1})$

Let $y = \ln a_n$ and $x = n$,

The linear equation is $y = x \ln 2 + \ln 5$ with the slope: $\ln 2$ and y-intercept: $\ln 5$.

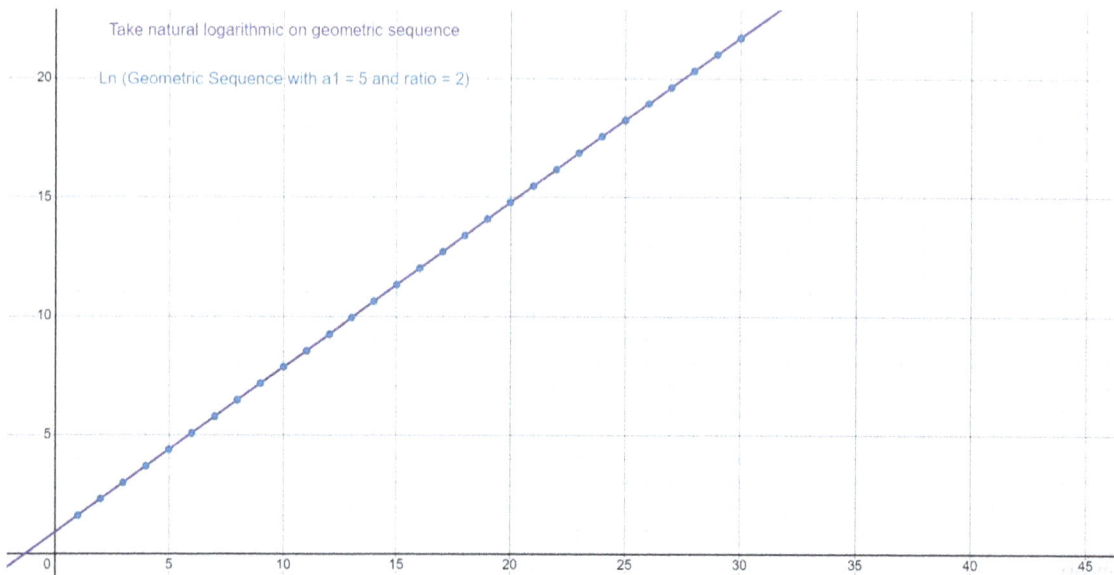

Take natural logarithmic on geometric sequence

Ln (Geometric Sequence with a1 = 5 and ratio = 2)

28

Example b2:

$$a_n = G(n, 5, 0.5) = 5 \cdot \left(\frac{1}{2}\right)^{n-1}, n = \{1, 2, 3, \dots\}$$

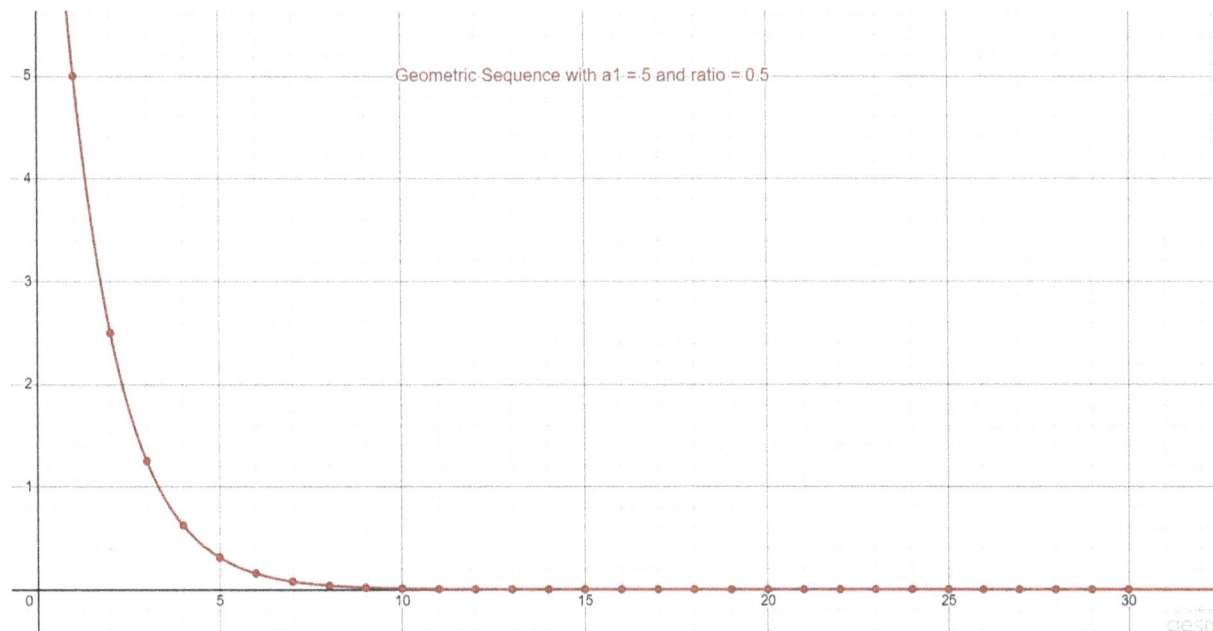

Taking logarithm: $\ln a_n = \ln(5 \cdot \left(\frac{1}{2}\right)^{n-1}) = \ln 5 \cdot 2^{-n+1}$

Let $y = \ln a_n$ and $x = n$,

The linear equation is $y = -x \ln 2 + \ln 2 + \ln 5$

$y = -x \ln 2 + \ln 10$ with the slope: $\ln 2$ and y-intercept: $\ln 10$.

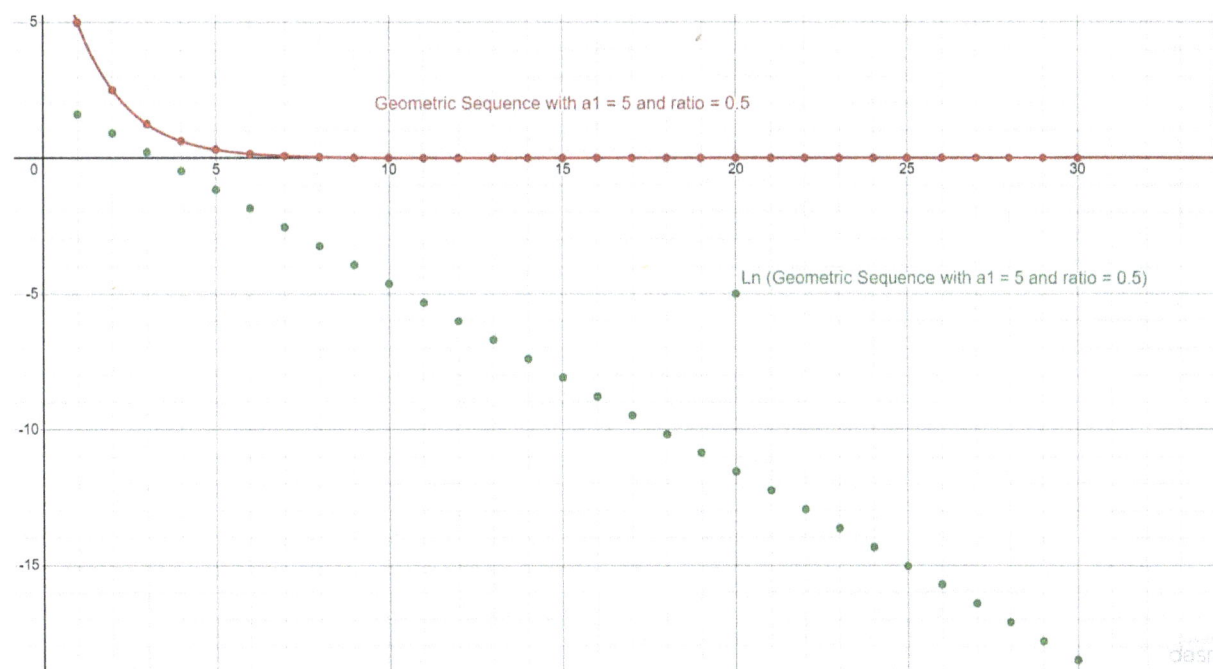

c. Fibonacci Sequences

$$F_1 = s_1, F_2 = s_2$$

$$F_n = F_{n-2} + F_{n-1}, n \geq 3$$

$$F_x = F(x) = F(x - 2) + F(x - 1), x = \{1, \dots, n, \dots\}$$

The graph shows the first 28 terms.

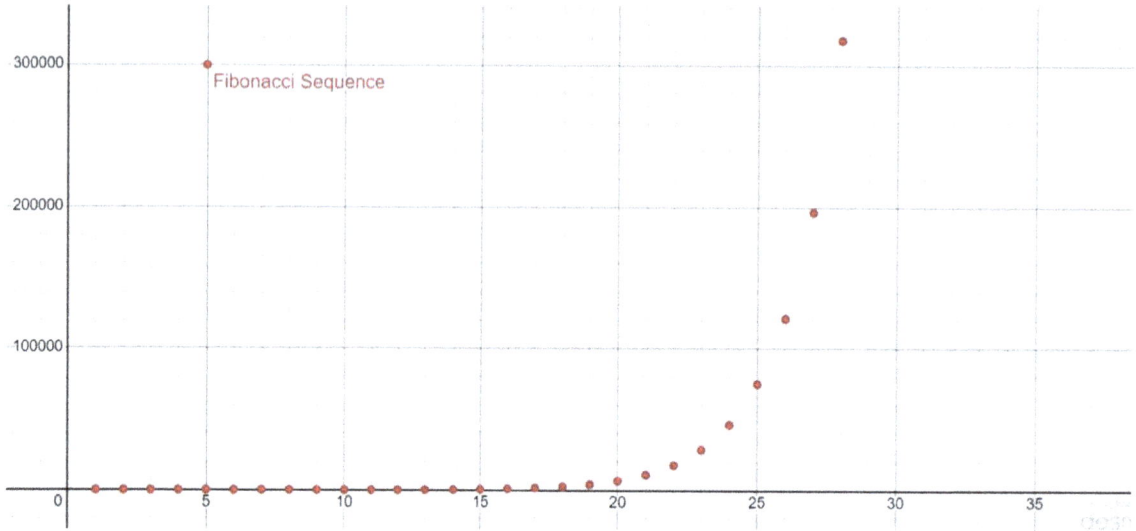

The graph shows the first 15 terms.

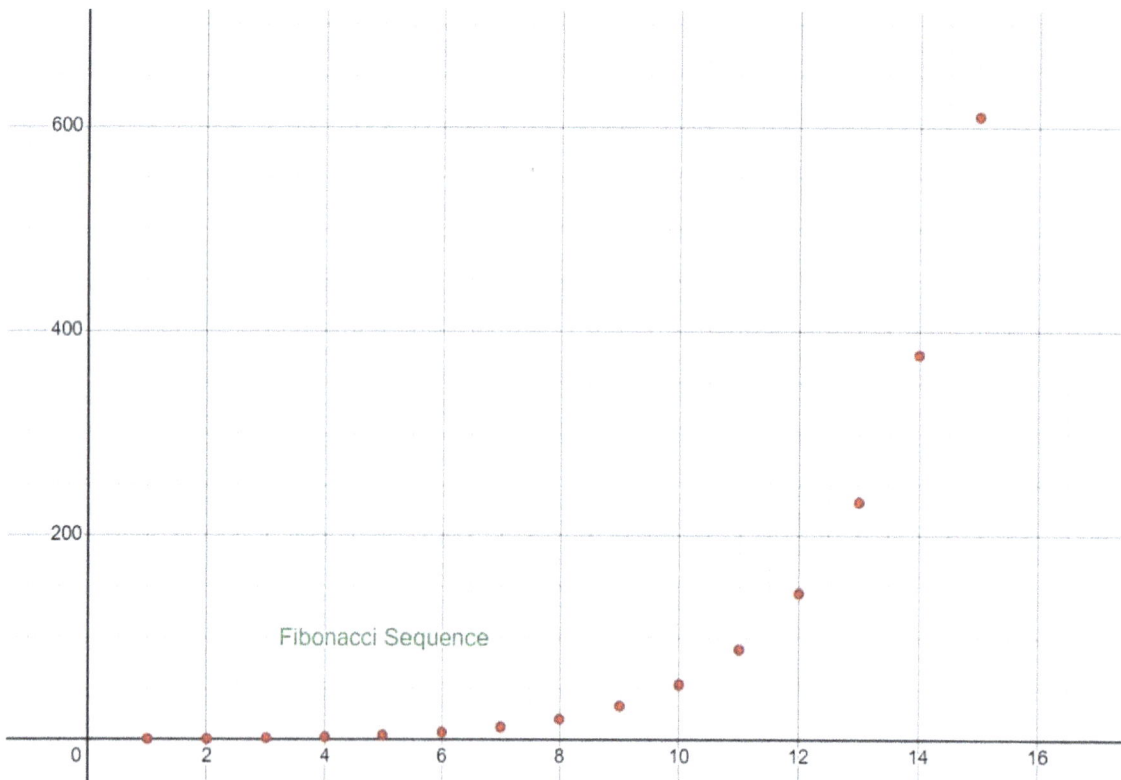

30

The graph shows the first 12 terms.

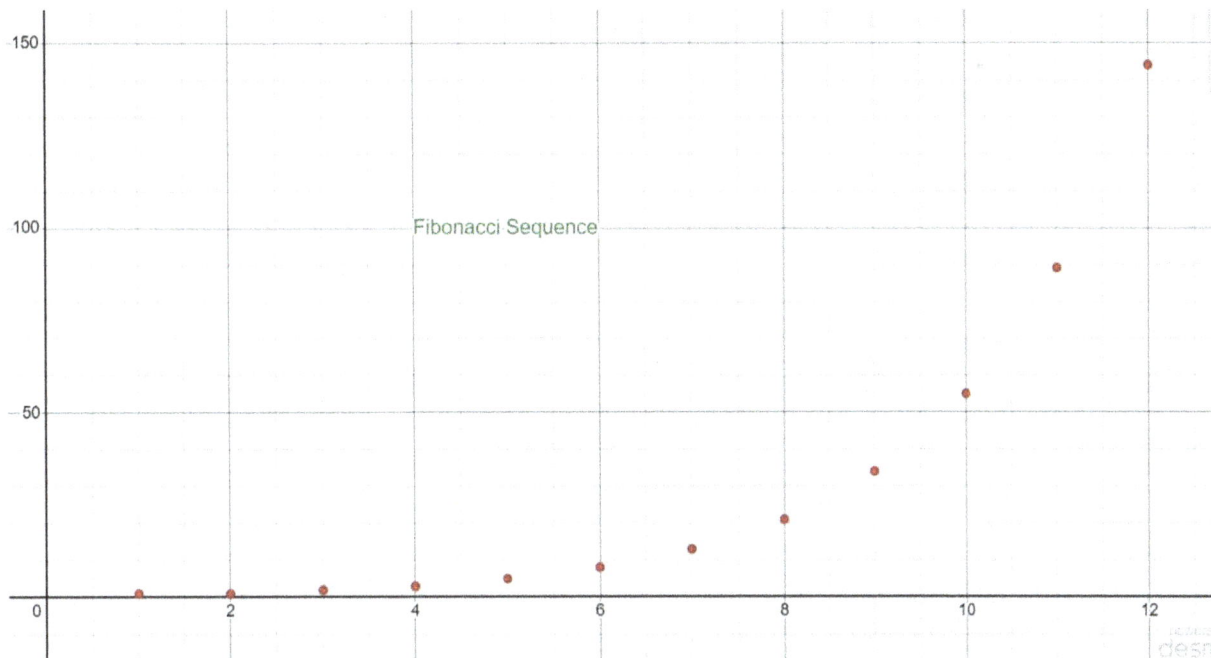

Taking logarithm: $\ln F(x) = \ln\big(F(x-2) + F(x-1)\big)$,

The linear equation is $y = 0.479852x + 0.776836$ with the slope of 0.479852 and y-intercept of 0.776836.

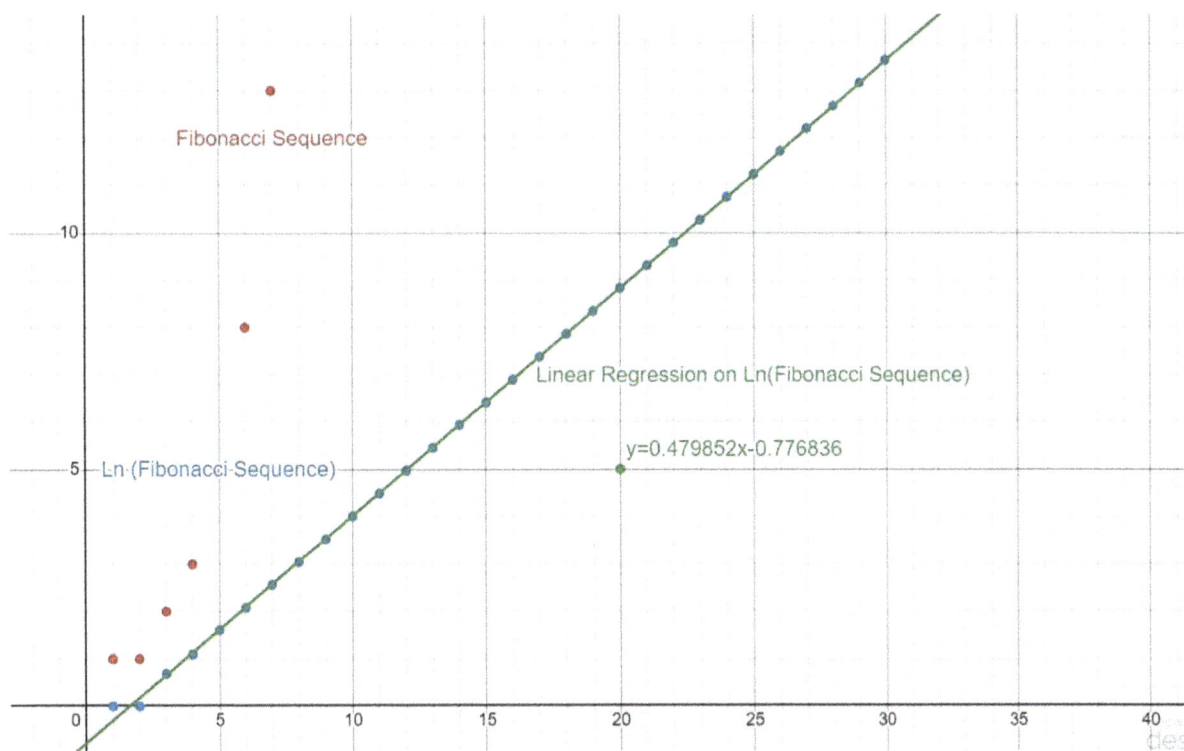

Another interesting fact about the Fibonacci sequence is the limit of the ratio of the nth term over (n-1) th term is the golden ratio number. That is $\lim\limits_{n \to \infty} \frac{F_n}{F_{n-1}} = \phi = \frac{1+\sqrt{5}}{2}$. See the table below.

Position	Fibonacci Sequence	ln(F(i))	Ratio
1	1	0	
2	1	0	1
3	2	0.693147	2
4	3	1.098612	1.5
5	5	1.609438	1.666666667
6	8	2.079442	1.6
7	13	2.564949	1.625
8	21	3.044522	1.615384615
9	34	3.526361	1.619047619
10	55	4.007333	1.617647059
11	89	4.488636	1.618181818
12	144	4.969813	1.617977528
13	233	5.451038	1.618055556
14	377	5.932245	1.618025751
15	610	6.413459	1.618037135
16	987	6.89467	1.618032787
17	1597	7.375882	1.618034448
18	2584	7.857094	1.618033813
19	4181	8.338306	1.618034056
20	6765	8.819518	1.618033963
21	10946	9.300729	1.618033999
22	17711	9.781941	1.618033985
23	28657	10.26315	1.61803399
24	46368	10.74436	1.618033988
25	75025	11.22558	1.618033989
26	121393	11.70679	1.618033989
27	196418	12.188	1.618033989
28	317811	12.66921	1.618033989
29	514229	13.15042	1.618033989
30	832040	13.63164	1.618033989

Golden Ratio Number
1.618033989

The golden ratio number $\frac{1+\sqrt{5}}{2}$ is the positive root of the quadratic equation:

$$x^2 - x - 1 = 0$$

d. Other Sequences

Figurate Numbers: Example Triangular Numbers

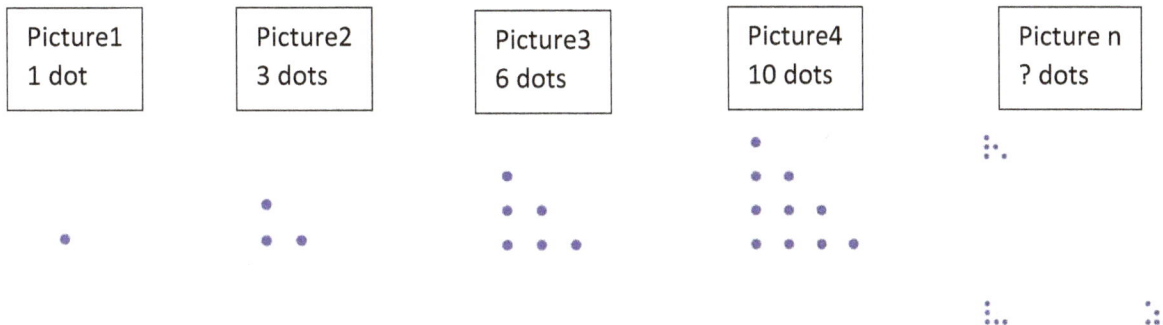

Picture1	Picture2	Picture3	Picture4	Picture n
1 dot	3 dots	6 dots	10 dots	? dots

We add the first n natural numbers to find how many dots there are in the nth picture. That is

$$1 + 2 + 3 + \cdots + n = \frac{(n+1)n}{2}, \ n = \{1,2,3,\dots\}$$

The above expression is called the arithmetic series.

The nth term formula is $a_n = \frac{(n+1)n}{2}$.

The first ten terms are
$\{1, 3, 6, 10, 15, 21, 28, 36, 45, 55\}$.

The table to the right shows the first 16 terms in the sequence.

The sequence is not an arithmetic sequence. Its first difference is an arithmetic. Its second difference is a constant. Its third difference is zero.

It is a discrete quadratic function.

n	nth term	first difference	Second difference
1	1		
2	3	2	
3	6	3	1
4	10	4	1
5	15	5	1
6	21	6	1
7	28	7	1
8	36	8	1
9	45	9	1
10	55	10	1
11	66	11	1
12	78	12	1
13	91	13	1
14	105	14	1
15	120	15	1
16	136	16	1

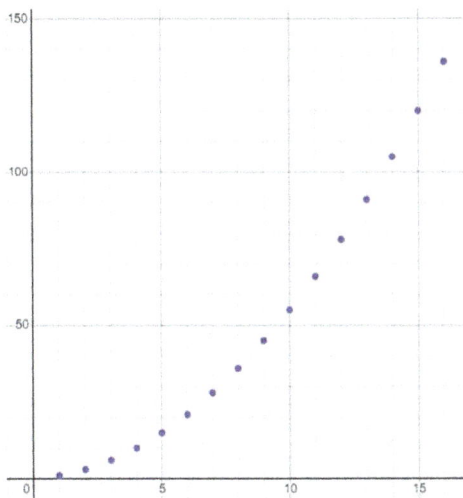

II.   Continuous Functions

The domain is a subset of the set of real numbers. We call the set of real numbers the support, environmental, universal, or ambient sets of a continuous function.

Examples:

1. Polynomial Functions
2. Normal Distribution
3. Student's t-Distribution
4. Chi-Square Distribution
5. F Distribution
6. Gamma Function
7. Gamma Distribution

II. Continuous Functions

1. Polynomial Functions

$$P(x, n, a_n, a_{n-1}, \ldots, a_0) = a_n x^n + a_{n-1} x^{n-1} + a_{n-2} x^{n-2} + \cdots + a_1 x + a_0,$$

$$x = (-\infty, \infty)$$

The following graphs are the five polynomial functions:

1. $P_0(x) = 10$, it is a constant function, and its degree is zero.
2. $P_1(x) = 2x + 5$, it is a linear function, and its degree is one.
3. $P_2(x) = 2x^2 + 30x + 108$, it is a quadratic function, and its degree is two.
4. $P_3(x) = x^3 + x^2 - 6x$, it is a cubic function, and its degree is three.
5. $P_4(x) = x^4 - 30x^3 + 323x^2 - 1470x + 2376$, it is a Quartic function, and its degree is four.

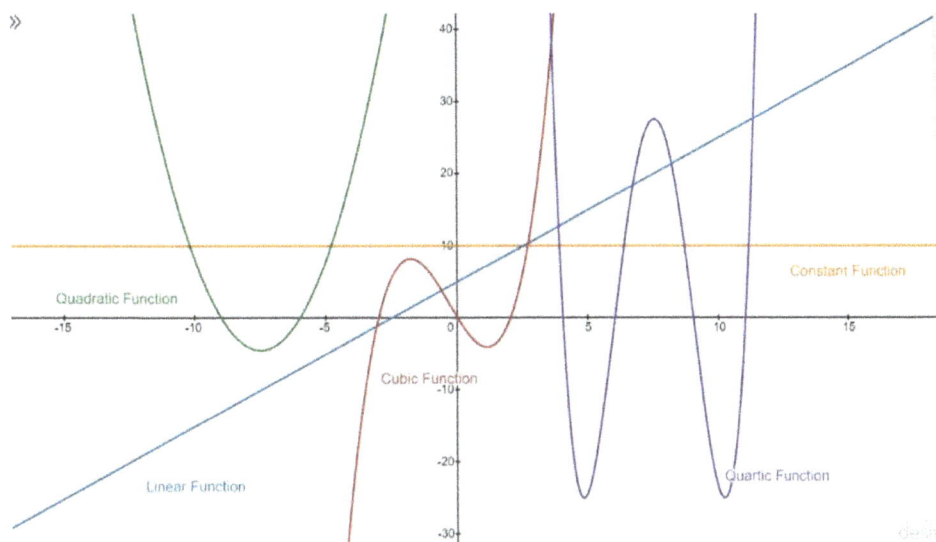

Find the x-intercepts of the above three functions.

$P_0(x)$ does not have x-intercept.

To find the x-intercepts of $P_1(x)$:

Solve $P_1(x) = 0$,

$$2x + 5 = 0$$

The solution is $x = -\frac{5}{2}$.

$P_1(x)$ has a x-intercept. That is $\left(-\frac{5}{2}, 0\right)$.

To find the x-intercepts of $P_2(x)$

Solve $P_2(x) = 0$,

$$2x^2 + 30x + 108 = 0$$

Factoring:

$$2(x^2 + 15x + 54) = 0$$

$$2(x + 9)(x + 6) = 0$$

The solutions are $x = -9 \; or -6$.

$P_2(x)$ has two x-intercepts.

They are $(-9, 0) \; and \; (-6, 0)$.

2. Normal Distribution

$$\phi(x, \mu, \sigma) = \frac{1}{\sigma\sqrt{2\pi}} \cdot e^{-\frac{(x-\mu)^2}{2\sigma^2}}, x = (-\infty, \infty)$$

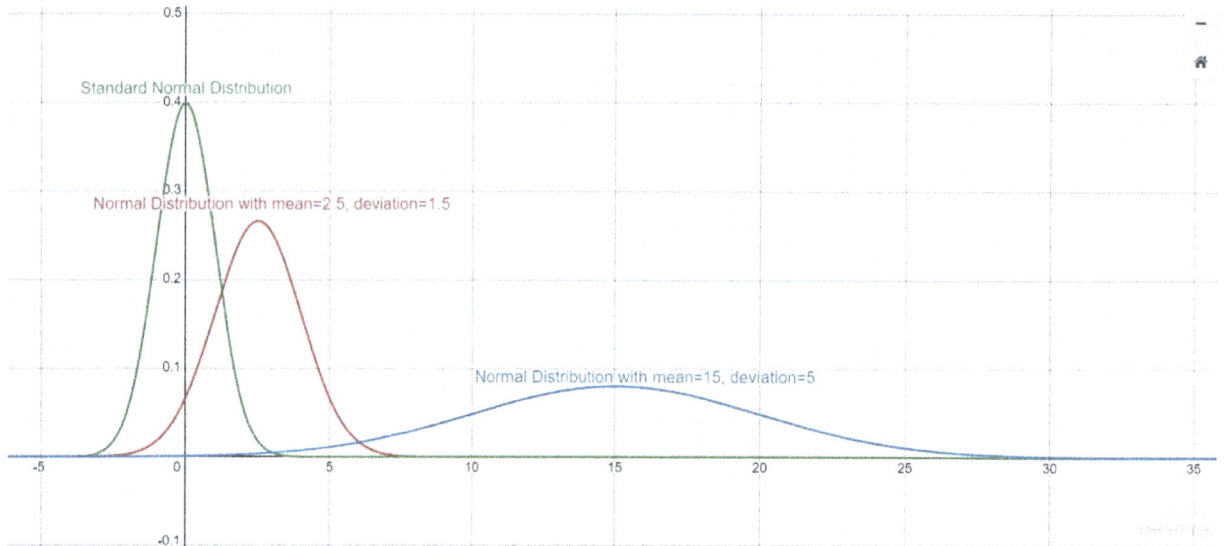

3. Student's t-Distribution

$$T(x, m) = \frac{k}{\left(1+\frac{x^2}{m}\right)^{\frac{m+1}{2}}}, k = \frac{\int_0^\infty x^{\frac{m+1}{2}-1} \cdot e^{-x} dx}{\sqrt{m\pi}\int_0^\infty x^{\frac{m}{2}-1} \cdot e^{-x} dx}, x = (-\infty, \infty), \text{ m is the degree of}$$

freedom.

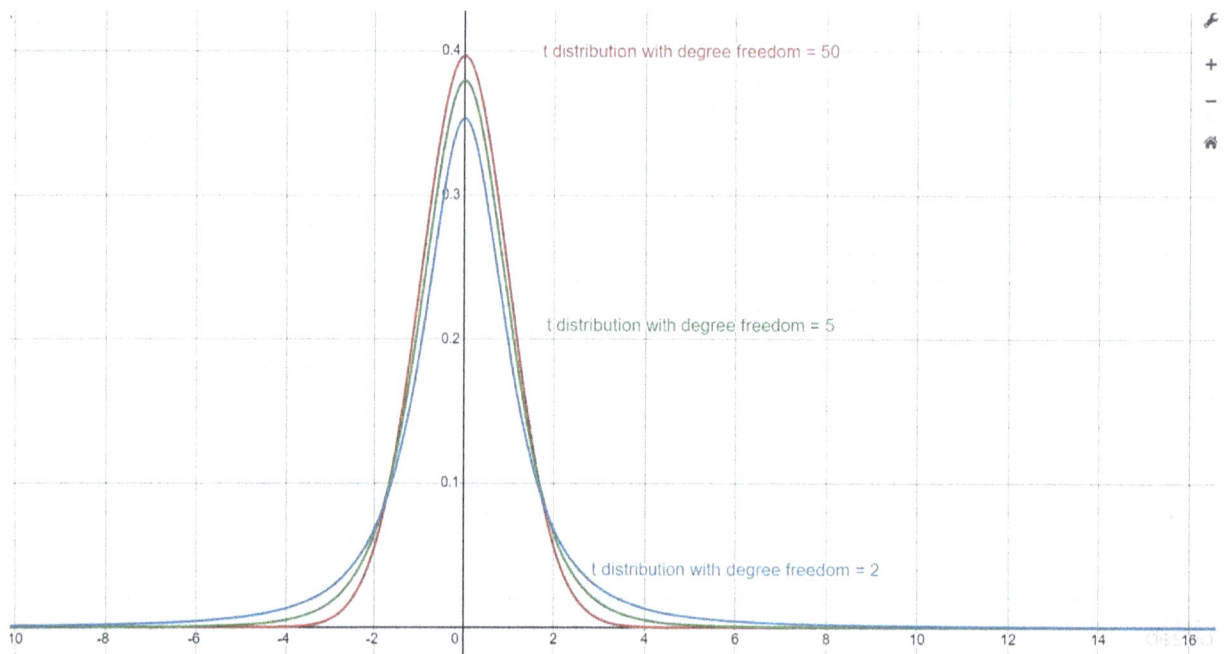

4. Chi-Square Distribution

$$\chi(x, n) = \frac{0.5^{\frac{n}{2}} x^{\frac{n}{2}-1} e^{-0.5x}}{\Gamma\left(\frac{n}{2}\right)}, x > 0$$

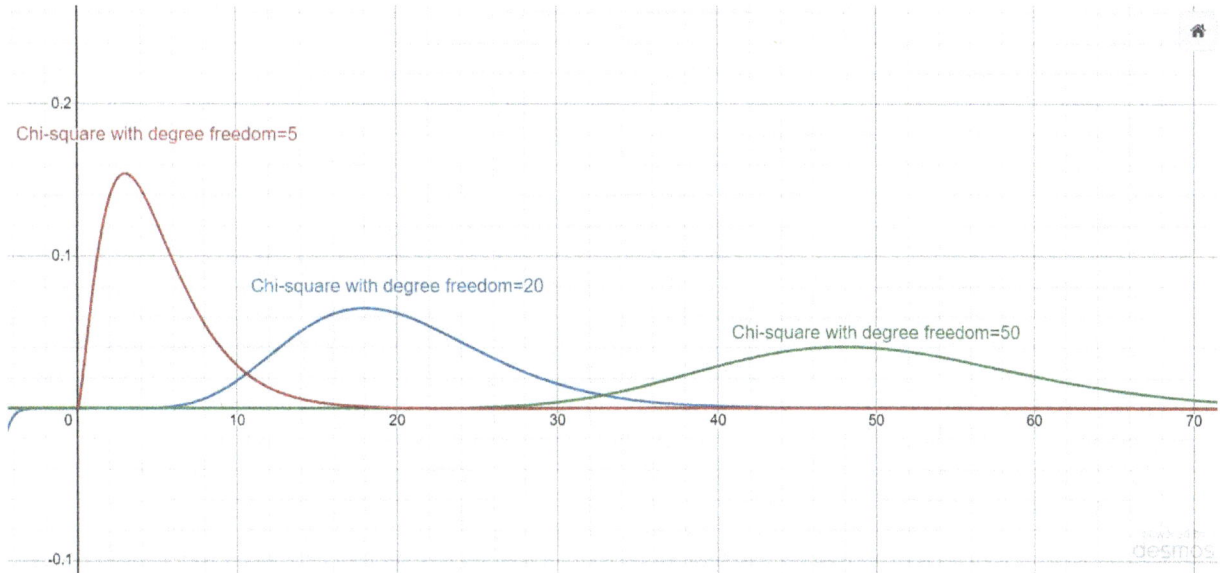

5. F Distribution

$$F(x, n, m) = \frac{\Gamma\left(\frac{n+m}{2}\right) \cdot n^{\frac{n}{2}} \cdot m^{\frac{m}{2}} \cdot x^{\frac{n}{2}-1}}{\Gamma\left(\frac{n}{2}\right) \cdot \Gamma\left(\frac{m}{2}\right) \cdot (m+nx)^{\frac{n+m}{2}}}, x > 0,$$ n is degree freedom one, and m is degree freedom 2.

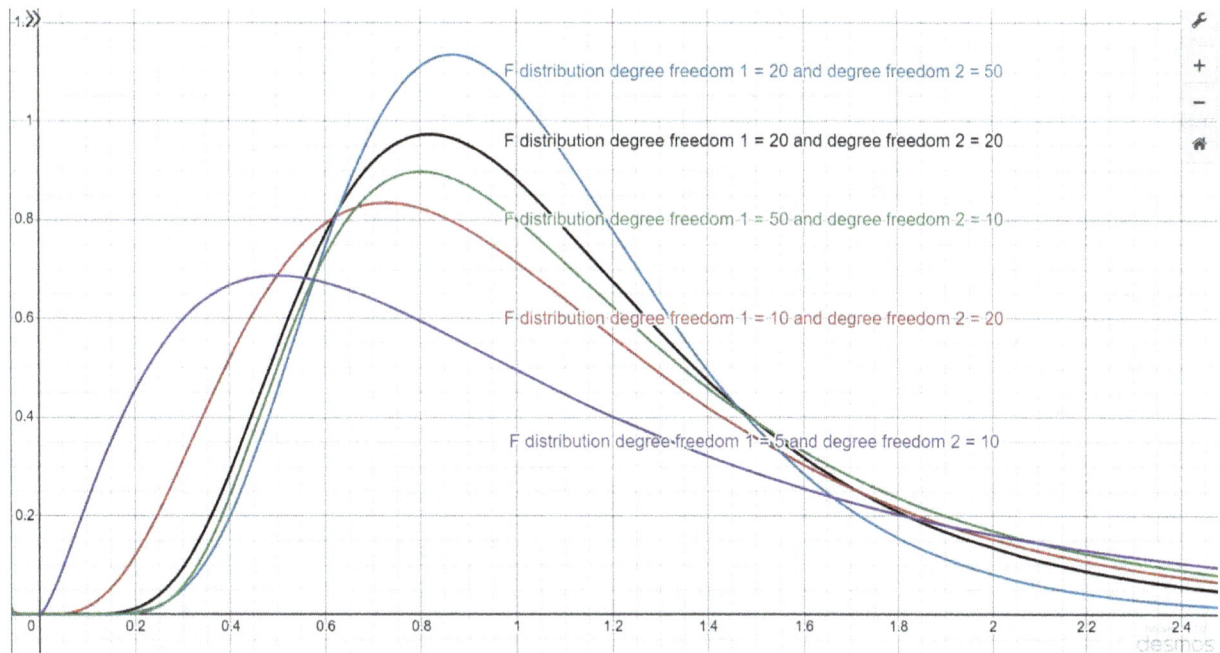

6. Gamma Function (Student's t, Chi-Square, F distributions using gamma function)

$$g(x) = \Gamma(x) = \int_0^\infty t^{(x-1)} \cdot e^{-t} dt, x > 0 \ \{x \ is \ shape\}$$

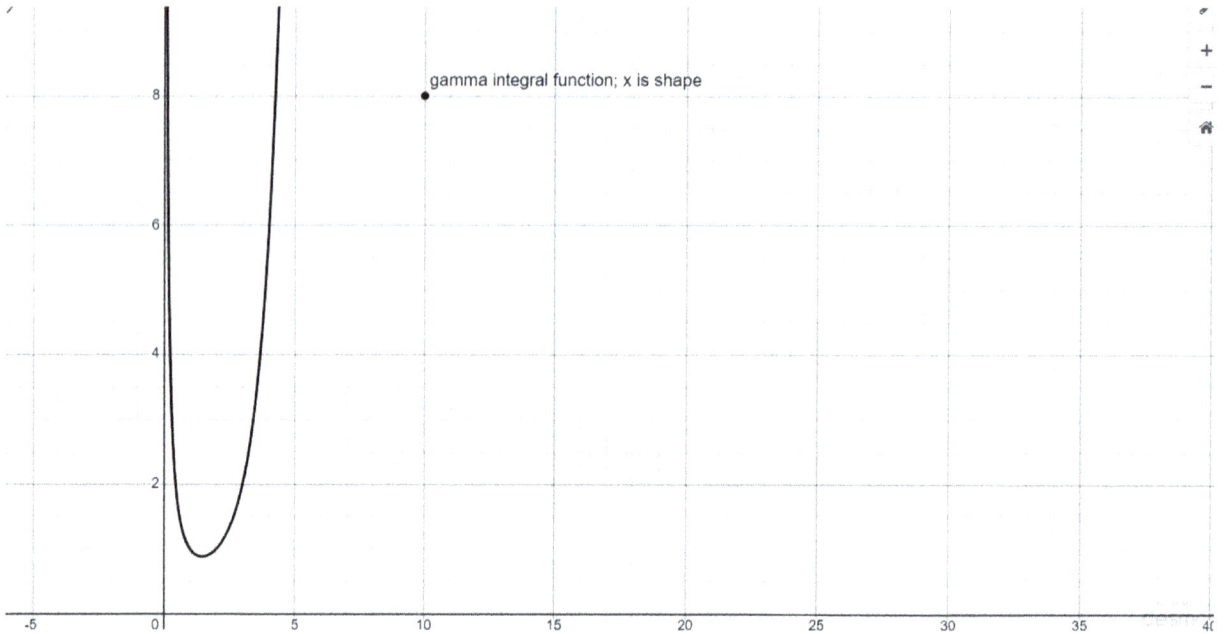

gamma integral function; x is shape

7. Gamma Distribution

$$G(x, r, \lambda) = \frac{\lambda^r x^{r-1} e^{-\lambda}}{\Gamma(r)}, x > 0$$

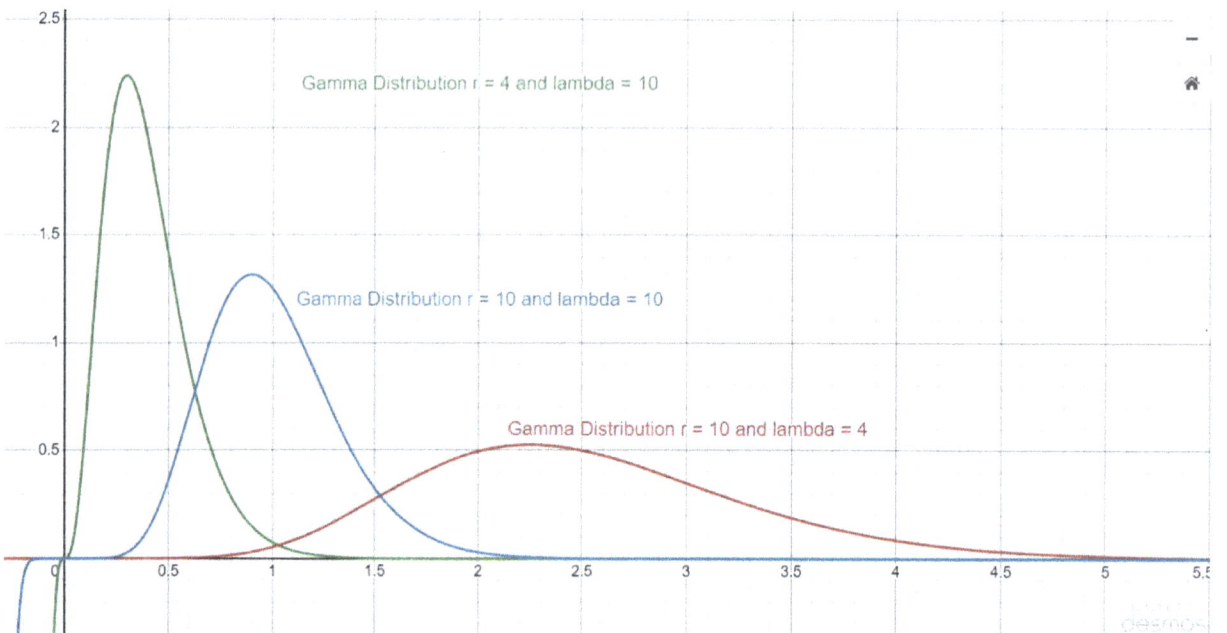

Gamma Distribution r = 4 and lambda = 10

Gamma Distribution r = 10 and lambda = 10

Gamma Distribution r = 10 and lambda = 4

38

III.   Algebraic Functions

Examples:

1. Polynomial Functions
2. Rational Functions
3. Cauchy Distribution

III.   Algebraic Functions

1.   Polynomial Functions

We define a polynomial function by its coefficients, degree, how many variables, and the domain. Let's define the domain as the set of real numbers. Then, variable values are real numbers with three algebraic operations: addition, subtraction, and multiplication. The degree is a natural number. We can limit the coefficients, such as the sets of whole numbers, integers, rationals, or real numbers.

Give an example with two variables, x and y, six terms (six coefficients), and a degree of 7.

$$F(x,y) = 4x^2y^5 - 3x^4y + 20x^5y + 8y^5 - 6x^6 + 5$$

The maximum of the sum of exponents of variables in each term is 7. The coefficients are 4, -3, 20, 8, -6, and 5, integers.

We define a general expression for a polynomial function with one variable and a degree of n as

$$P(x) = a_nx^n + a_{n-1}x^{n-1} + a_{n-2}x^{n-2} + \cdots + a_1x + a_0$$

Where the coefficients are $a_n, a_{n-1}, a_{n-2}, \ldots, a_1, a_0$.

In the real function analysis, the domain of a function is the set of real numbers.

({$P(x)$}, +, *) has an algebraic Ring structure.

(R, +, *) has an algebraic field structure and $a_i \in R$.

Polynomials have many applications. One of the applications of polynomials is to form decimals in any base numeration system, such as the base-ten system. We can store the form of decimals as a polynomial; the $P(x)$ to $P(b, n, A)$, where b is the base, n is the number of digits, and A is the list containing all digits.

Example-1: How does a computer read 123,546 (one-hundred twenty-three thousand and five hundred forty-six)?

The computer will set

x = 10, n = 6, A = [1, 2, 3, 5, 4, 6]

$$P(x, n, A) = \sum_{i=1}^n A[i]x^{n-i}$$

Compute

$P(10, 6, A) = 123546$; see figure 6

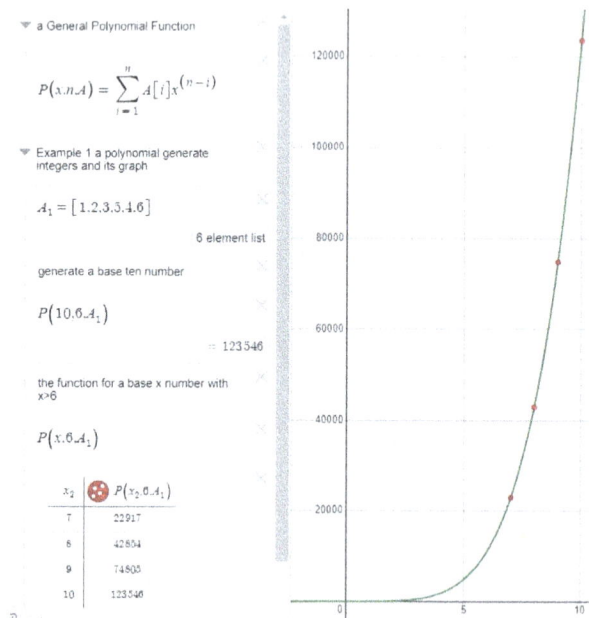

▼ a General Polynomial Function

$$P(x, n, A) = \sum_{i=1}^n A[i]x^{(n-i)}$$

▼ Example 1 a polynomial generate integers and its graph

$A_1 = [1, 2, 3, 5, 4, 6]$

6 element list

generate a base ten number

$P(10, 6, A_1)$

= 123546

the function for a base x number with x>6

$P(x, 6, A_1)$

x_2	$P(x_2, 6, A_1)$
7	22917
8	42854
9	74805
10	123546

Let us look at a base-sixteen number. Base-sixteen has sixteen digits: {0, 1, 3, …9, a, b, c, d, e, f}

Example 2: Input the number 5a2bed4c6 into a computer and convert it to base-ten.

<table>
<tr><td>

The computer will set

x = 16, n = 9,

A = [5, a, 2, b, e, d, 4, c, 6]

  = [5,10,2,11,14,13,4,12,6]

$$P(x,n,A) = \sum_{i=1}^{n} A[i]x^{n-i}$$

Compute

$$P(16,9,A) = 2.4205251782 \times 10^{10}$$

see figure 7. The graph with logarithmic on y-Axis.

</td><td>

a General Polynomial Function

$$P(x,n,A) = \sum_{i=1}^{n} A[i]x^{(n-i)}$$

Example 2 a polynomial generate integers in bases 15,16   and its graph

$$A_1 = [5,10,2,11,14,13,4,12,6]$$

9 element list

convert the base sixteen number to a base ten

$$P(16,9,A_1)$$

$$= 2.4205251782 \times 10^{10}$$

the function for a base x number with x>14

$$P(x,9,A_1)$$

x_2	$P(x_2,9,A_1)$
15	1.463483×10^{10}
16	2.420528×10^{10}
17	3.90473×10^{10}
20	1.409655×10^{11}

</td></tr>
</table>

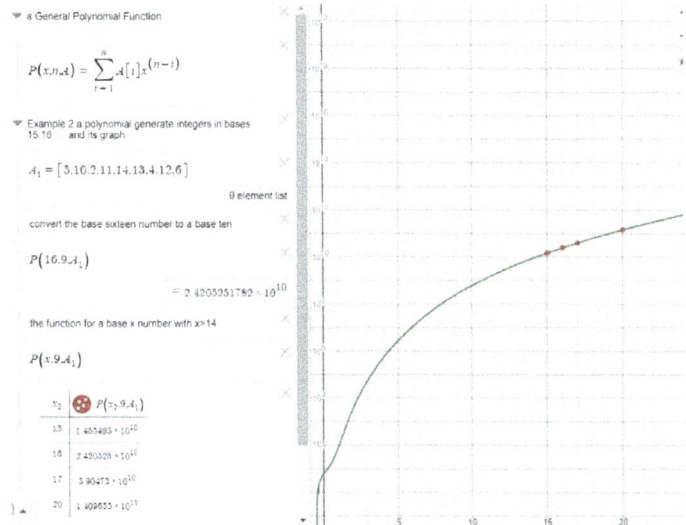

Example 3: Generate the finite decimal 321.0546 (three-hundred twenty-one and five hundred forty-six ten-thousandths. P is the polynomial Function with the independent variable for the number of digits to the right of the decimal point.

<table>
<tr><td>

The computer will set

x = 10, n = 7, m = 4,

A = [3, 2, 1, 0, 5, 4, 6]

$$P(x,n,m,A) = 10^{-m}\sum_{i=1}^{n} A[i]x^{n-i}$$

Compute

$$P(10,7,4\ A) = 321.0546$$

see figure 7.

The graph a polynomial function of m.

</td><td>

a General Polynomial Function

$$P(x,n,m,A) = 10^{-m}\sum_{i=1}^{N} A[i]x^{(n-i)}$$

Example 3: a polynomial generate decimals in base-ten and its graph

$$A_1 = [3,2,1,0,5,4,6]$$

7 element list

generate the decimal

$$P(10,7,4,A_1)$$

$$= 321.0546$$

the function for base-ten decimals

$$P(10,7,x,A_1)$$

x_2	$P(10,7,x_2,A_1)$
4	321.0546
5	32.10546
6	3.210546
7	0.3210546

</td></tr>
</table>

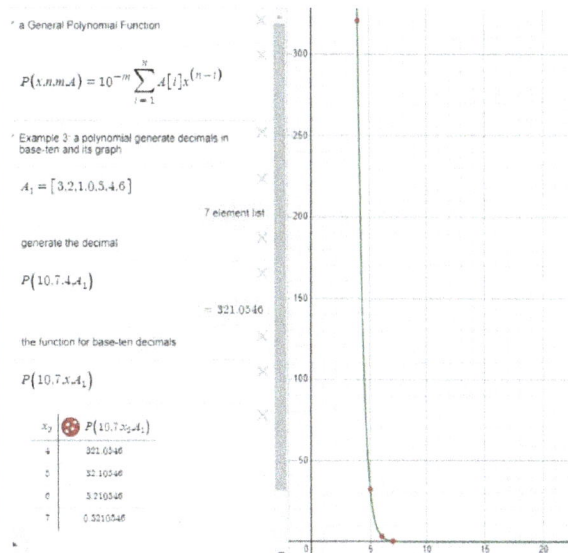

2. Rational Functions $Q(x)$

The quotient of two polynomials defines a rational function.

Variable values are real numbers that exclude the denominator's zeros and have four algebraic operations: addition, subtraction, Multiplication, and Division. When the denominator is a number, the rational Function is a polynomial.

$(\{Q(x)\}, +, *)$ has an algebraic Field structure. $(R, +, *)$ has an algebraic field structure and $a_i, b_i \in R$.

$$R(x) = \frac{a_n x^n + a_{n-1} x^{n-1} + a_{n-2} x^{n-2} + \cdots + a_1 x + a_0}{b_m x^m + b_{m-1} x^{m-1} + b_{m-2} x^{m-2} + \cdots + b_1 x + b_0} = \frac{p(x)}{q(x)}, p(x) \text{ and } q(x) \text{ are polynomials,}$$

Domain: $\{x | b_m x^m + b_{m-1} x^{m-1} + b_{m-2} x^{m-2} + \cdots + b_1 x + b_0 \neq 0 \text{ and } x \in (-\infty, \infty)\}$

Example 1: rational functions with one variable

<table>
<tr>
<td>

a. the graph of

$$f(x) = \frac{-6x^6 + 20x^5 - 3x^4 + 4x^2 + 13}{-6x^5 - 2x^4 + 20x^3 + 4x^2 + 25}$$

The function has a vertical asymptote:

$x = a$ and $a \in (1.90916, 1.90917)$

The function has a slate asymptote:

$$y = x - \frac{22}{6}$$

</td>
<td>

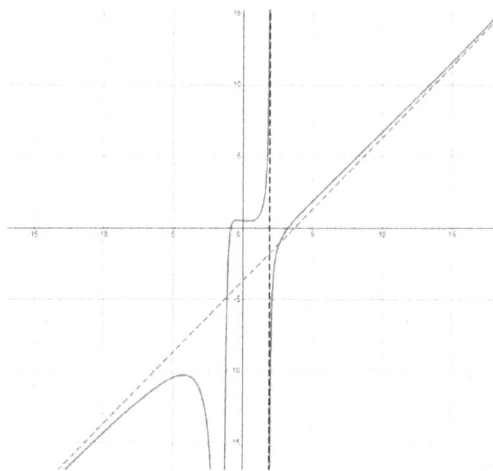

</td>
</tr>
<tr>
<td>

b. the graph of

$F(x) = f(x, g(x))$ with

$$f(x, g(x)) = \frac{(4x^2+8)(g(x))^5 - (3x^4 - 20x^5)g(x) - 6x^6 + 5}{(5x+8)(g(x))^5 - (2x^4 - 20x^3)g(x) - 6x^5 + 15}$$

$g(x) = x^2 - x - 1$

The function has three vertical asymptotes:

$x = [a_1, a_2, a_3]$, $a_1 \in (-1.03095, -1.03096)$

$a_2 \in (0.1329, 0.13295), a_3 \in (2.06111, 2.06112)$

The function has a slate asymptote:

$$y = 0.8x - 1.28$$

</td>
<td>

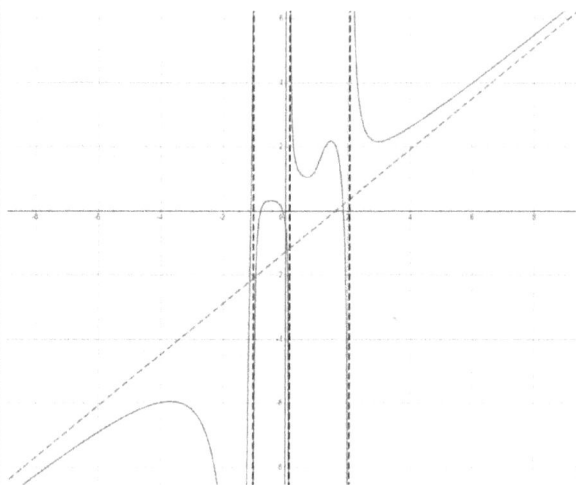

</td>
</tr>
</table>

Example 2: a polynomial function with two variables x and y

$$F(x,y) = \frac{4x^2y^5 - 3x^4y + 20x^5y + 8y^5 - 6x^6 + 5}{5xy^5 - 2x^4y + 20x^3y + 8y^5 - 6x^5 + 15}$$

The numerator is $p(x,y) = 4x^2y^5 - 3x^4y + 20x^5y + 8y^5 - 6x^6 + 5$, and

the denominator is $q(x,y) = 5xy^5 - 2x^4y + 20x^3y + 8y^5 - 6x^5 + 15$

Then $F(x,y) = \frac{p(x,y)}{q(x,y)}$

The graph of $F(x,y)$ in 3D

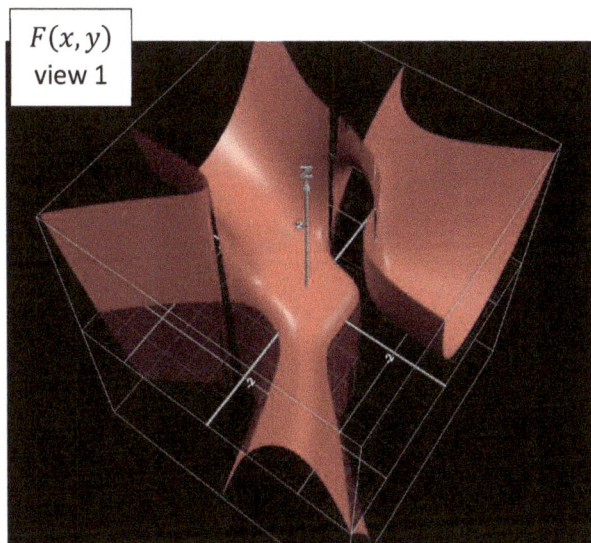

$F(x,y)$ view 1

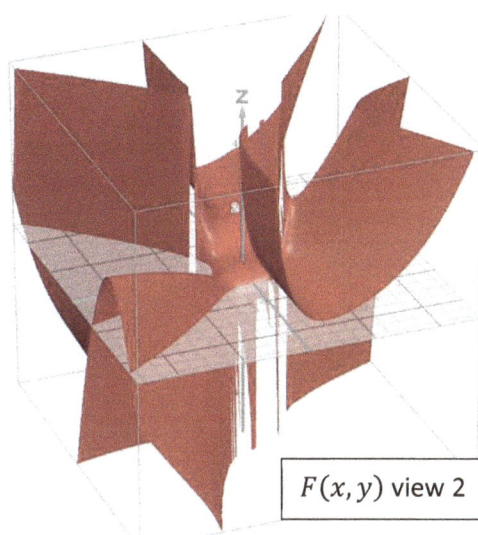

$F(x,y)$ view 2

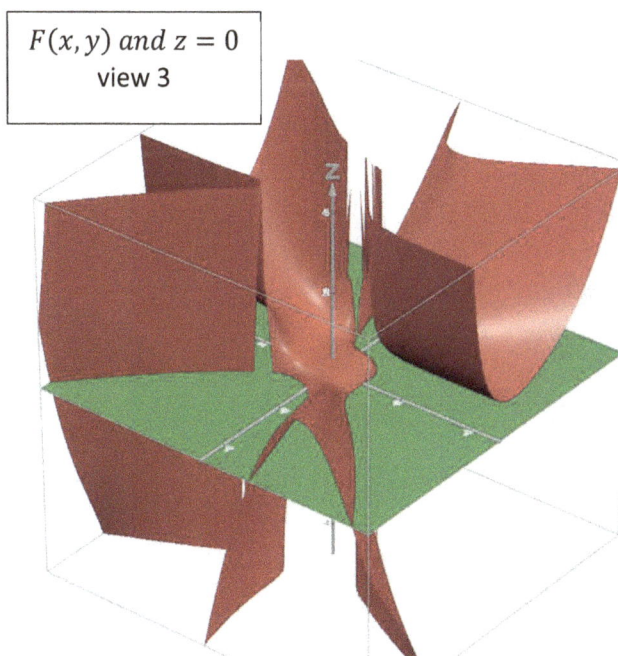

$F(x,y) \ and \ z = 0$
view 3

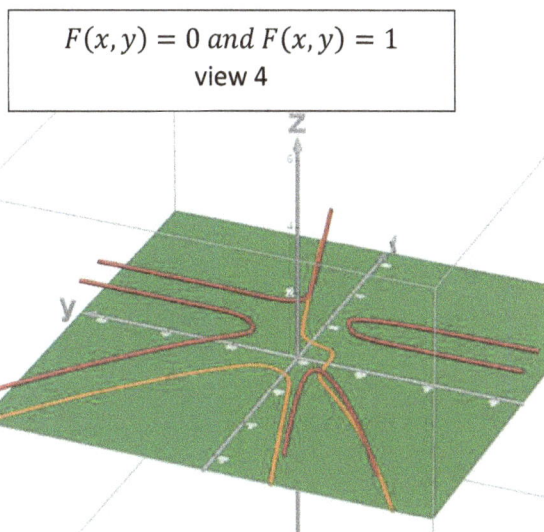

$F(x,y) = 0 \ and \ F(x,y) = 1$
view 4

3. Cauchy Distribution

$$f(x) = \frac{1}{\pi(1 + x^2)}, x = (-\infty, \infty)$$

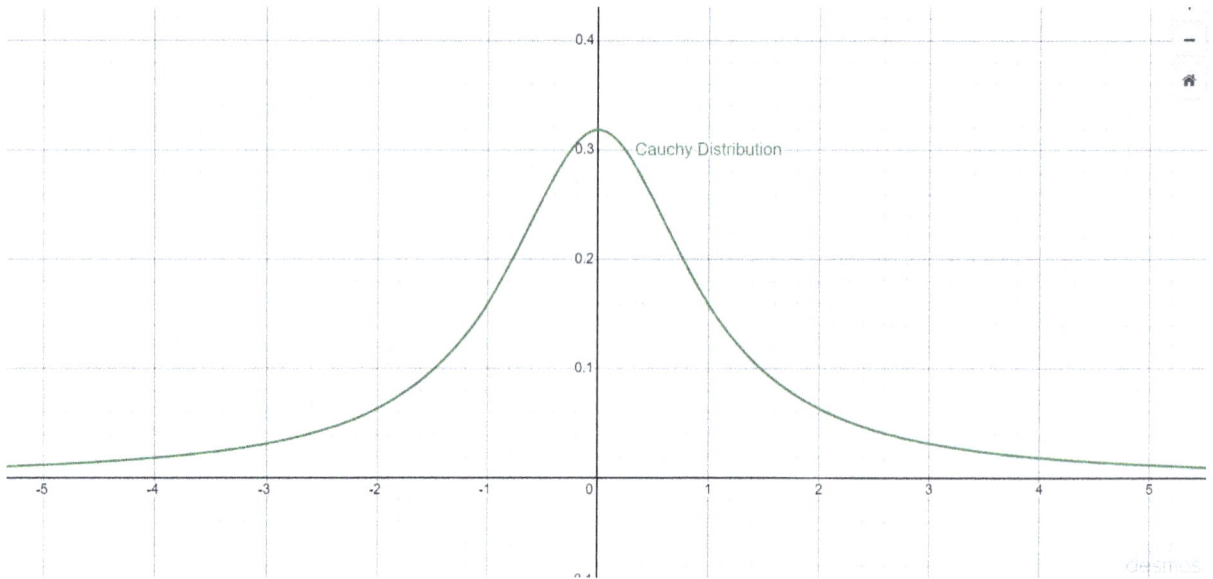

$$\int_{-\infty}^{\infty} \frac{1}{\pi(1 + x^2)} dx = 1$$

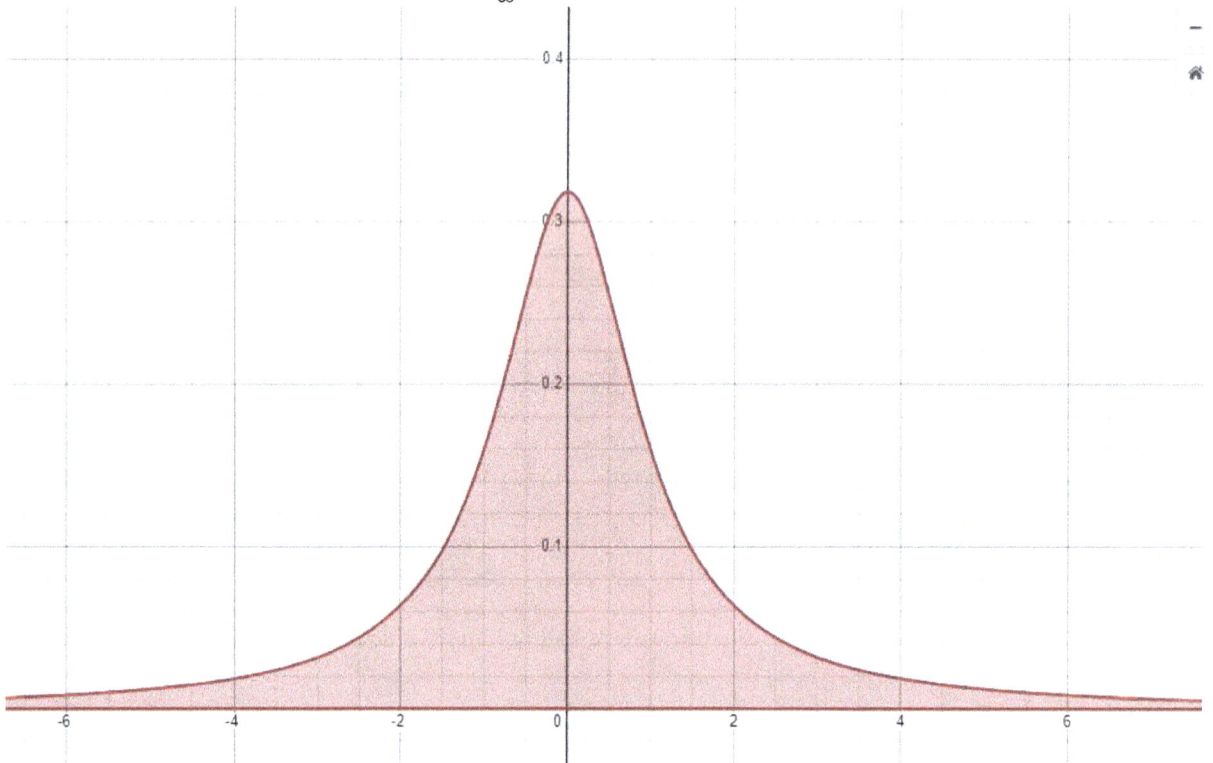

IV.   Transcendental Functions

Examples:

1. Trigonometric Functions
    i.   Six primary trigonometric functions
    ii.   Six primary inverse trigonometric functions
    iii. Combining algebraic functions and trigonometric functions
    iv.
2. Exponential Functions
    i.   Primary exponential functions
    ii.   Transformations of primary exponential functions
    iii. Combine algebraic and exponential functions
3. Logarithmic Functions
    i.   Primary logarithmic functions
    ii.   Transformations of the primary logarithmic functions
    iii. Combine algebraic and logarithmic functions

1. Trigonometric Functions
   i. Six primary trigonometric functions

Trigonometric functions have a rich historical development dating back to ancient civilizations such as the Babylonians, Egyptians, and Greeks. The Babylonians were among the first to study angles and ratios in triangles, laying the foundation for the development of trigonometry. The Egyptians also used trigonometric concepts in their architectural and surveying practices. However, the Greeks, particularly Hipparchus and Ptolemy, formalized the study of trigonometry by introducing the trigonometric functions we recognize today, such as sine, cosine, and tangent. These functions became essential tools in astronomy, navigation, and mathematics, eventually leading to the sophisticated trigonometric relationships and identities we use extensively in various fields today.

Definitions:

The foundation is given a right triangle with labels as the figure to the right:

$$0° < x < 90° \ in \ degree \ or \ 0 < x < \frac{\pi}{2} \ in \ radians$$

1. The Sine Relation is $\sin(x) = \frac{a}{c}$
2. The Cosine Relation is $\cos(x) = \frac{b}{c}$
3. The Tangent Relation is $\tan(x) = \frac{a}{b}$
4. The Cotangent Relation is $\cot(x) = \frac{b}{a}$
5. The Secant Relation is $\sec(x) = \frac{c}{b}$
6. The Cosecant Relation is $\csc(x) = \frac{c}{a}$

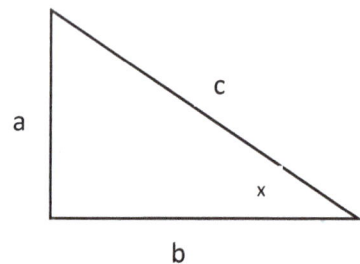

The transition in the historical development of trigonometric functions from the context of right triangles to the unit circle and eventually to analytical functions marked a significant advancement in mathematical understanding. While trigonometry initially focused on relationships within right triangles, the shift toward the unit circle expanded its applications and provided a more general framework. The unit circle allowed for a deeper exploration of trigonometric concepts beyond acute angles, enabling the development of sine and cosine functions as coordinates on the unit circle. This shift paved the way for developing trigonometric identities and extending trigonometric functions to complex numbers. The six trigonometric functions integrate into the category of analytical functions that play a crucial role in modern mathematics and its applications.

Definitions: six analytical trigonometric functions

(the bridge between the geometry and analytical functions)

The foundation is given the unit circle centered at the origin $(0, 0)$.

θ is the central angle in radians. S is the traveling distance from the initial point $(1,0)$ to the terminal point (x,y) with directions, count clockwise is positive and clockwise is negative. And using the relationship between the central angle and the corresponding arc length such as $S = r\theta$. Since $r = 1$, so $S = \theta$.

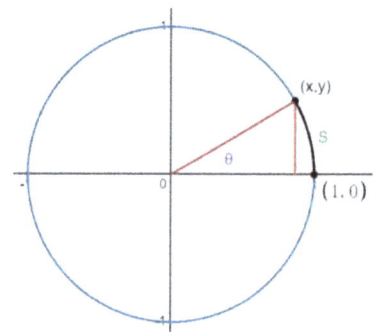

Six primary trigonometric functions are

1. The sine function $f(\theta) = \sin(\theta) = \sin(S) = f(S) = y$
2. The cosine function $f(\theta) = cos(\theta) = \cos(S) = f(S) = x$
3. The tangent Function $f(\theta) = \tan(\theta) = \tan(S) = f(S) = \frac{y}{x}$
4. The cotangent function $f(\theta) = \cot(\theta) = \cot(S) = f(S) = \frac{x}{y}$
5. The secant function $f(\theta) = \sec(\theta) = \sec(S) = f(S) = \frac{1}{x}$
6. The cosecant function $f(\theta) = \csc(\theta) = \csc(S) = f(S) = \frac{1}{y}$

We can extend the universal set/ambient set and codomain from the above definitions to the set of real numbers. Now, we clearly understand the domain and range for each trigonometric Function. See the table below.

Now, we let x be the input, the independent variable, and Function notation be the output, the dependent variable. We can express the six primary trigonometric functions as the following:

d. $f(x) = \sin(x)$ e. $f(x) = \cos(x)$ f. $f(x) = \tan(x)$	a. $f(x) = \csc(x)$ b. $f(x) = \sec(x)$ c. $f(x) = \cot(x)$

Trigonometric functions	Domain	Range
Sine $f(x) = \sin(x)$	The set of real numbers or $(-\infty, \infty)$	The set of real numbers from negative one to positive one inclusive or $[-1,1]$
Cosine $f(x) = \cos(x)$	The set of real numbers or $(-\infty, \infty)$	The set of real numbers from negative one to positive one inclusive or $[-1,1]$
Tangent $f(x) = \tan(x)$	The set of real numbers, excluding the odd multiples of $\frac{\pi}{2}$ or $\left\{x \middle\| x \neq \frac{(2n+1)\pi}{2}\right\}, n$ is any integer.	The set of real numbers or $(-\infty, \infty)$
Cotangent $f(x) = \cot(x)$	The set of real numbers excluding the integer multiples of π or $\{x\|x \neq n\pi\}, n$ is any integer.	The set of real numbers or $(-\infty, \infty)$
Secant $f(x) = \sec(x)$	The set of real numbers, excluding the odd multiples of $\frac{\pi}{2}$ or $\left\{x \middle\| x \neq \frac{(2n+1)\pi}{2}\right\}, n$ is any integer.	The set of real numbers excluding the real numbers from negative one to positive one exclusive or $(-\infty, -1] \cup [1, \infty]$
Cosecant $f(x) = \csc(x)$	The set of real numbers excluding the integer multiples of π or $\{x\|x \neq n\pi\}, n$ is any integer.	The set of real numbers excluding the real numbers from negative one to positive one exclusive or $(-\infty, -1] \cup [1, \infty]$

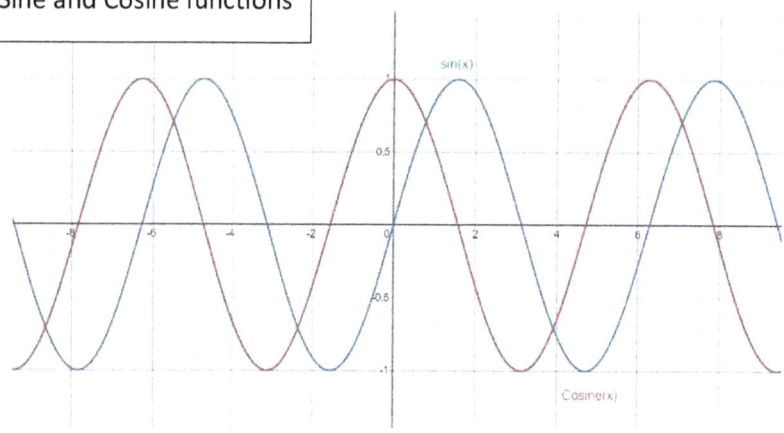
The graphs of Sine and Cosine functions

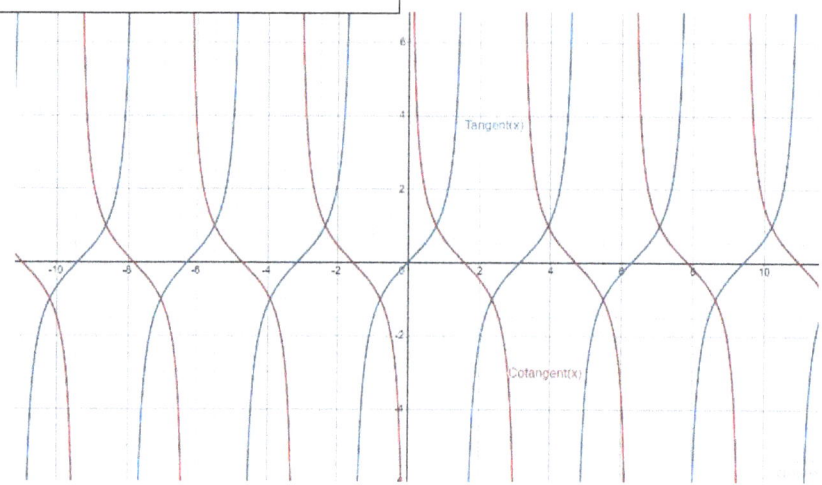
The graphs of Tangent and Cotangent functions

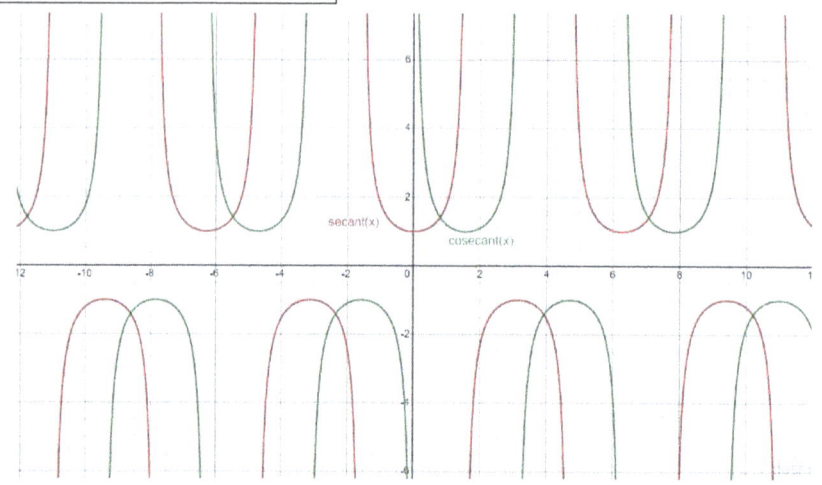
The graphs of Secant and Cosecant functions

48

ii. The six primary inverse trigonometric functions

a. $f(x) = \sin^{-1}(x)$
b. $f(x) = \cos^{-1}(x)$
c. $f(x) = \tan^{-1}(x)$
d. $f(x) = \cot^{-1}(x)$
e. $f(x) = \sec^{-1}(x)$
f. $f(x) = \csc^{-1}(x)$

Inverse trigonometric functions	Domain	Range
Arcsine	The set of real numbers from negative one to positive one inclusive or $[-1,1]$	The set of real numbers from negative half pi to positive half pi inclusive or $\left[-\frac{\pi}{2}, \frac{\pi}{2}\right]$
Arccosine	The set of real numbers from negative one to positive one inclusive or $[-1,1]$	The set of real numbers from zero to pi inclusive or $[0, \pi]$
Arctangent	The set of real numbers or $(-\infty, \infty)$	The set of real numbers from negative half pi to positive half pi exclusive or $(-\frac{\pi}{2}, \frac{\pi}{2})$
Arccotangent	The set of real numbers or $(-\infty, \infty)$	The set of real numbers from zero to pi exclusive or $(0, \pi)$
Arcsecant	The set of real numbers excluding the real numbers from negative one to positive one exclusive or $(-\infty, -1] \cup [1, \infty]$	The set of real numbers from negative half pi to positive half pi inclusive or $\left[-\frac{\pi}{2}, \frac{\pi}{2}\right]$
Arccosecant	The set of real numbers excluding the real numbers from negative one to positive one exclusive or $(-\infty, -1] \cup [1, \infty]$	The set of real numbers from zero to pi or $[0, \pi]$

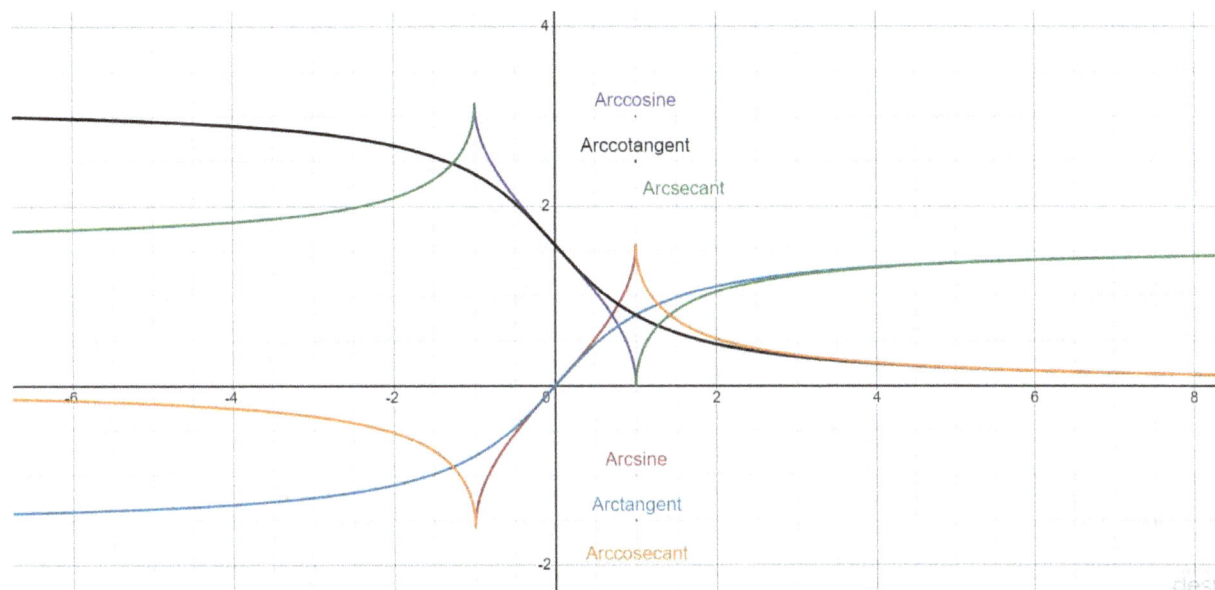

iii. Combining algebraic and trigonometric functions
Five operations/combinations on functions are +, -, *, /, °: addition, subtraction, multiplication, division, and composition.

Example1: Let a polynomial be $P(x) = 0.01(x + 9)(x + 5)(3x + 7)(x - 1)(2x + 3)$.

Study the pattern of two compositions:

a. The composition of the given polynomial and sine function
b. The composition of the sine function and the given polynomial

Use the two functions $P(x)$ and $S(x) = \sin(x)$

1. a The mathematical expression for the composition of the given polynomial and sine function is

$$C_1(x) = P(x)°S(x) = P(S(x))$$
$$= 0.01(\sin(x) + 9)(\sin(x) + 5)(3\sin(x) + 7)(\sin(x) - 1)(2\sin(x) + 3)$$

A graph of three functions together:

The graph of $C_1(x)$ is in purple color.

The graph of $\sin(x)$ is in red color.

The graph of $P(x)$ is in blue color.

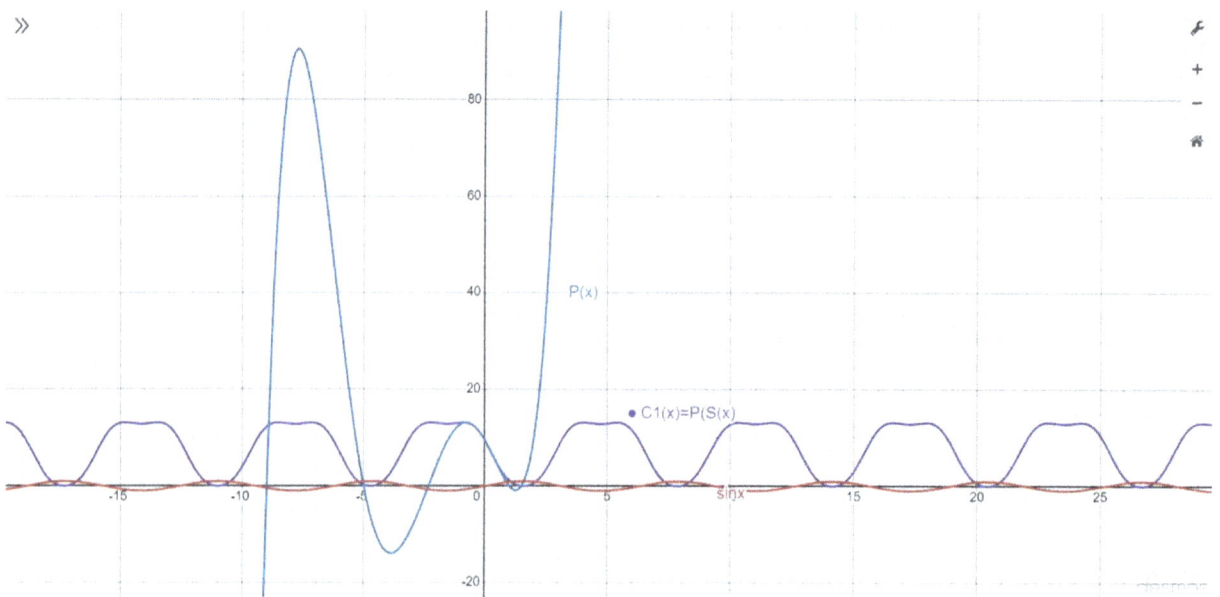

50

1. b The mathematical expression for the sine function and the composition of the given polynomial is

$$C_2(x) = S(x) \circ P(x) = S(P(x)) = \sin\big(0.01(x+9)(x+5)(3x+7)(x-1)(2x+3)\big)$$

A graph of three functions together:

The graph of $C_2(x)$ is in purple color.

The graph of $\sin(x)$ is in red color.

The graph of $P(x)$ is in blue color.

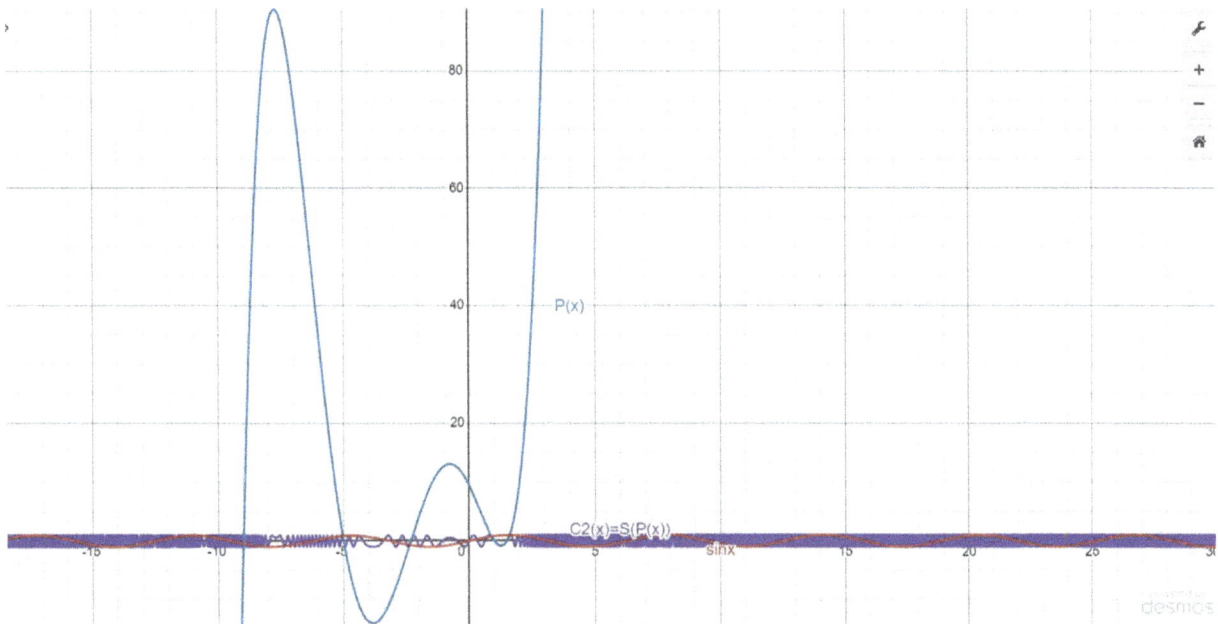

Exam the graph of $C_2(x)$ let y-axis = [-1.5,1.5]

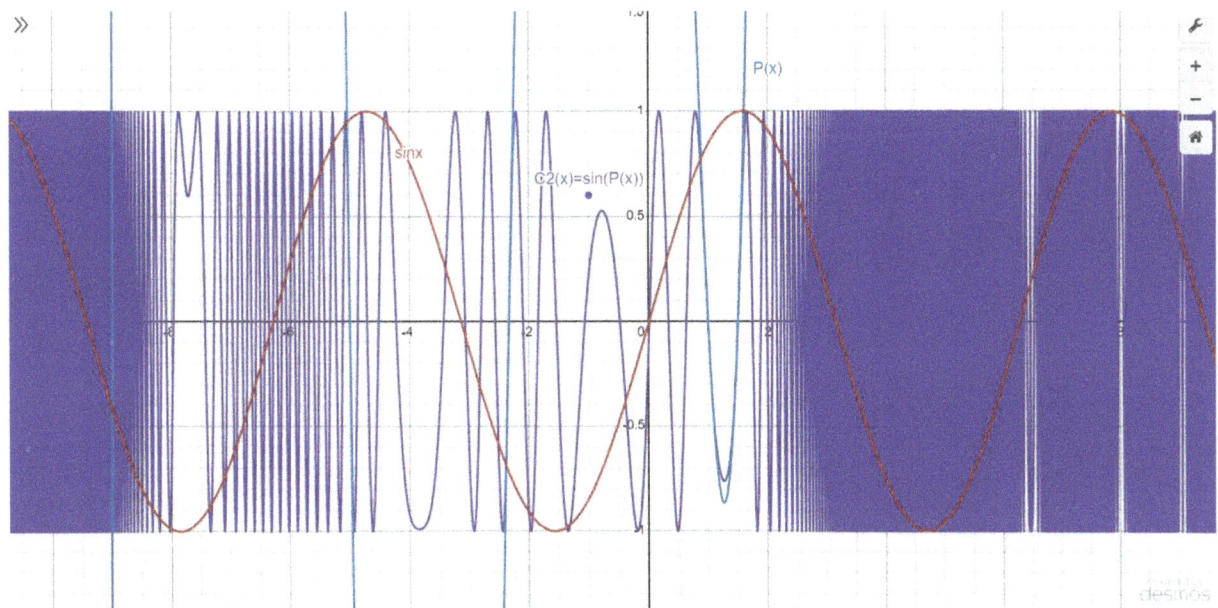

2. Exponential Functions
i. Primary exponential functions

$f(x) = b^x$, b is a positive real number excluding 1. That is $0 < b < 1 \; or \; b > 1$.

Primary exponential functions' domain is the set of real numbers, and range is the set of positive real numbers.

The graphs of three primary exponential functions: $f(x) = 1.5^x$, $f(x) = 2^x$, and $f(x) = 3^x$

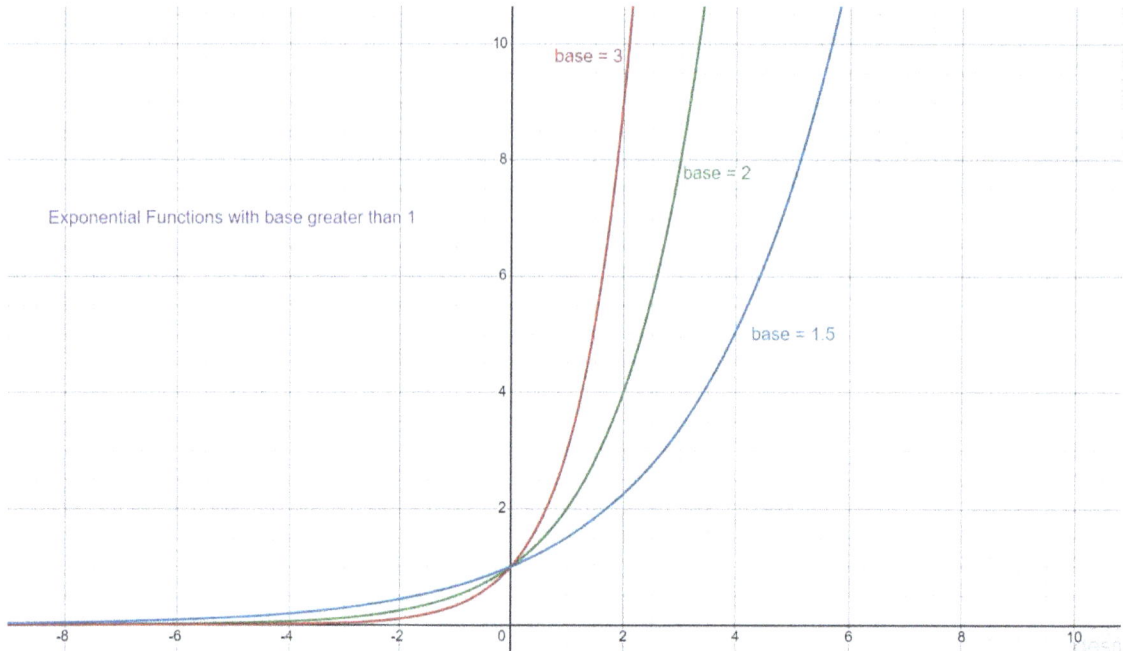

The graphs of three primary exponential functions: $f(x) = \left(\frac{1}{3}\right)^x$, $f(x) = \left(\frac{1}{2}\right)^x$, and $f(x) = \left(\frac{2}{3}\right)^x$

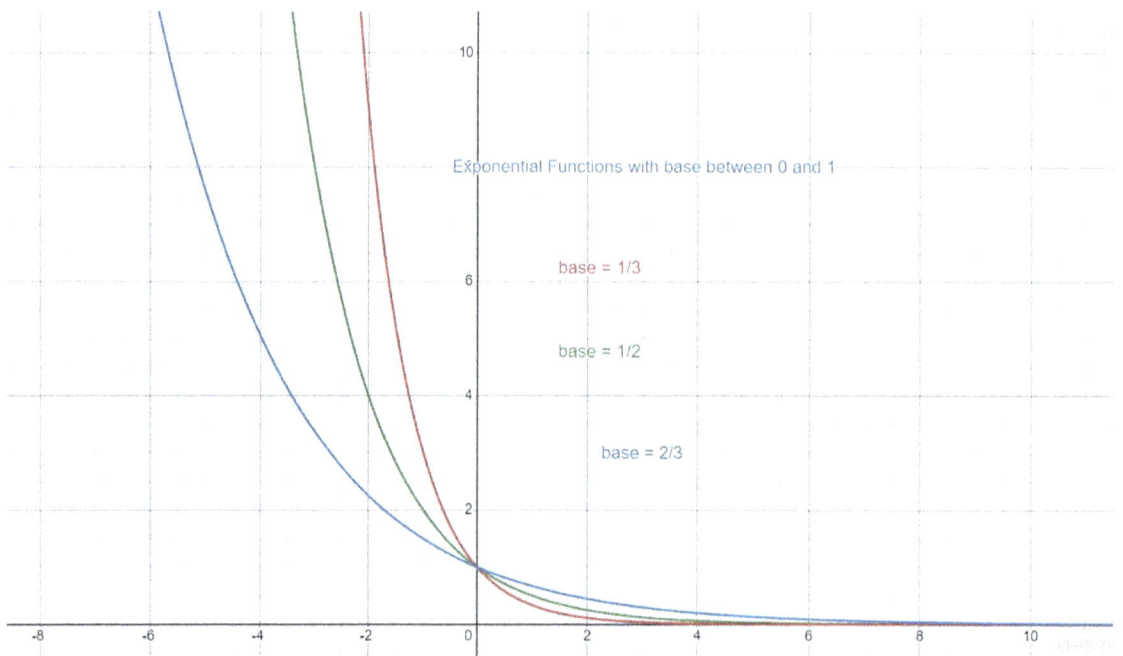

ii.   Transformations of primary exponential functions

There are three types of transformations: horizontal translation, vertical translation, and scaling, which include positive and negative real numbers. When a scaling is negative, we will interpret the scaling as a reflection and a scaling.

$f(x) = a \cdot b^{x-h} + k$, b is a positive real number excluding 1. That is $0 < b < 1 \; or \; b >$

Its domain is the set of real numbers; its range is all real numbers larger than k.

The scaler $a \neq 0$; $h, k \in \mathbb{R}$; $\mathbb{R}$ is the set of real numbers. h is the horizontal translation, and k is the vertical translation.

Let $a = 1, h = 5, k = -8$

The graph of $f(x) = 3^{x-5} - 8$ is in the green color.

The graph of $f(x) = 3^x$ is in the red color.

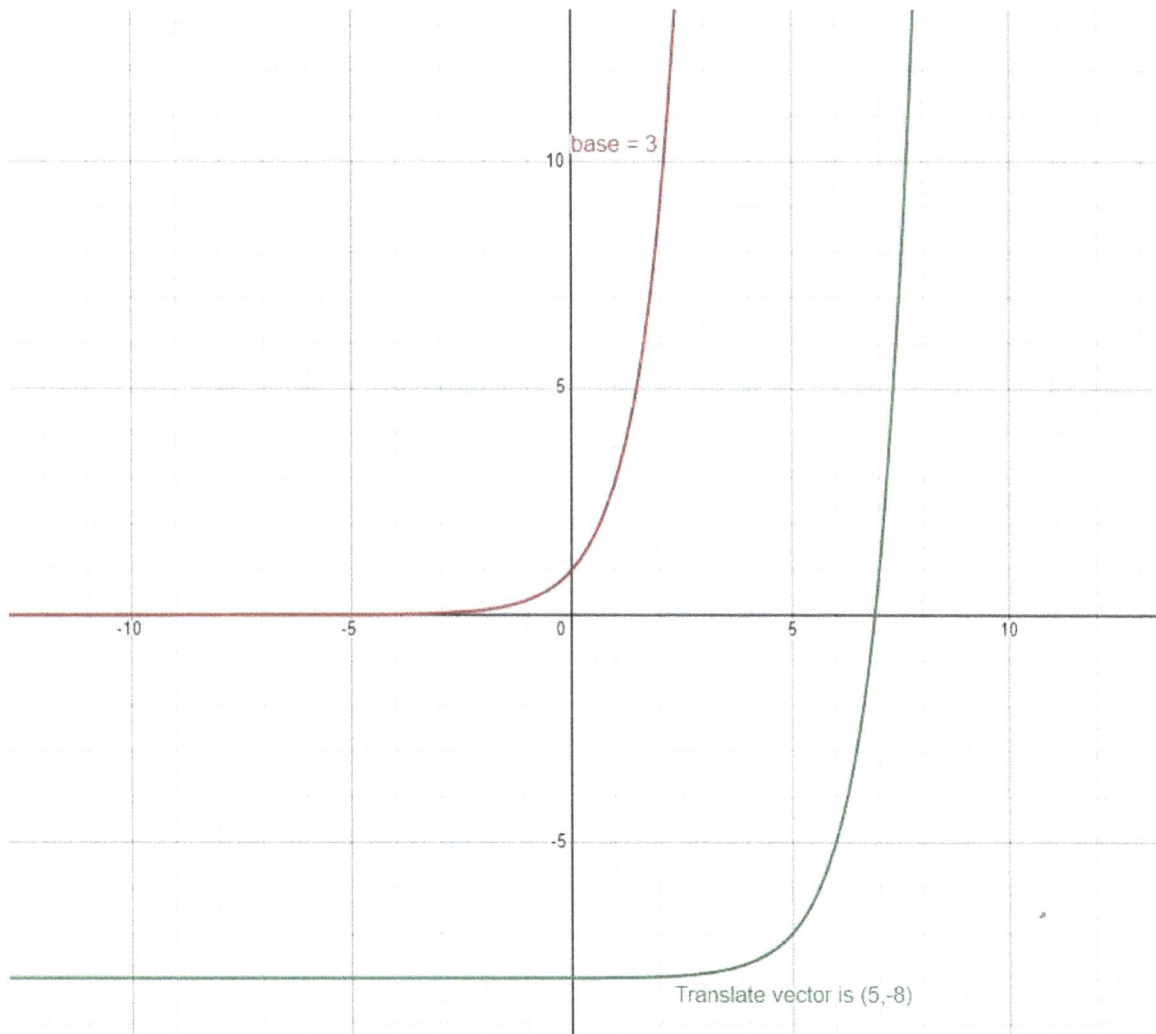

base = 3

Translate vector is (5,-8)

iii. Combining algebraic and exponential functions

Five operations/combines on functions are +, -, *, /, °: addition, subtraction, multiplication, division, and composition.

Example Rayleigh Distribution

$$R(x) = xe^{-\frac{1}{2}x^2}, x \geq 0$$

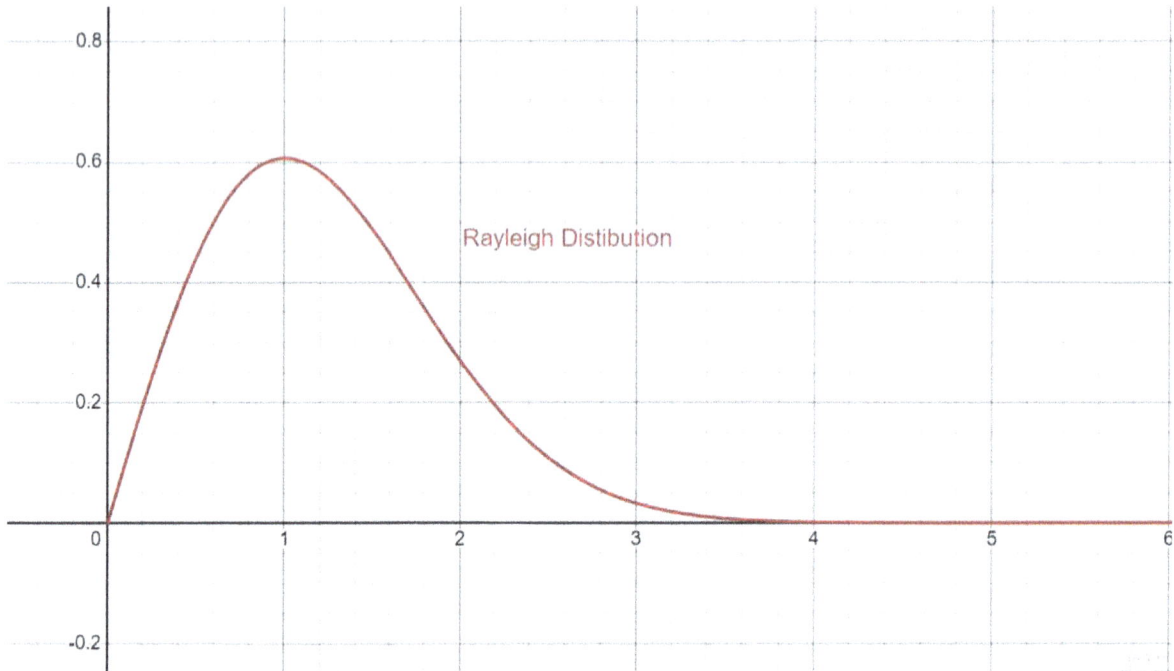

A decomposition of the Rayleigh Distribution function is the result of multiplication and composition of the three functions:

1. $f(x) = x, x \geq 0$, it is the identity function with nonnegative inputs.
2. $g(x) = \frac{1}{2}x^2$, it is a quadratic function.
3. $h(x) = e^x$, it is the primary natural exponential Function.

$$R(x) = f(x) \cdot h\big(g(x)\big)$$

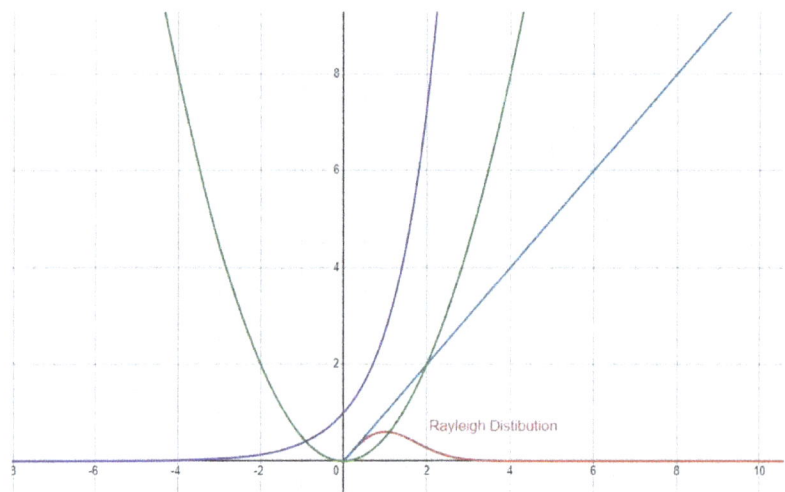

3. Logarithmic Functions
   i.   Primary logarithmic functions

$f(x) = \log_b x$, b is a positive real number excluding 1. That is $0 < b < 1\ or\ b > 1$.

Its domain is the set of positive real numbers, and range is the set of real numbers.

The graphs of three primary logarithmic functions with a base greater than 1:

$f(x) = \ln x$ in blue color,

$f(x) = \log_2 x$ in black color, and $f(x) = \log_5 x$ in red color.

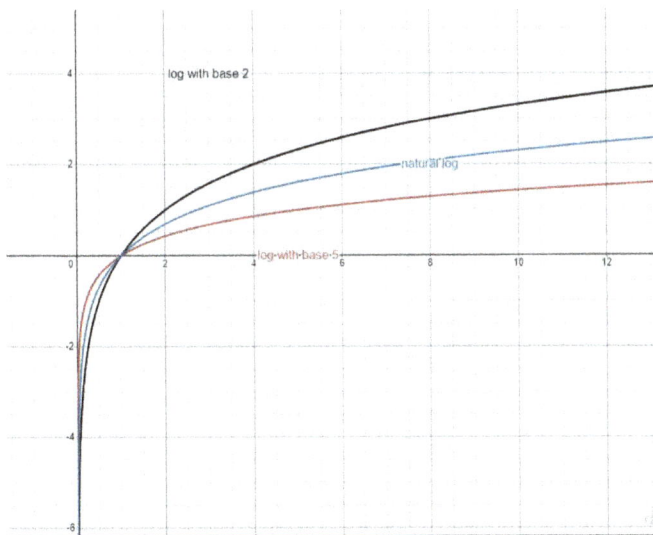

The graphs of three primary logarithmic functions with a base of less than 1:

$f(x) = \log_{\frac{1}{e}} x$ in blue color,

$f(x) = \log_{\frac{1}{2}} x$ in black color, and $f(x) = \log_{\frac{1}{5}} x$ in red color.

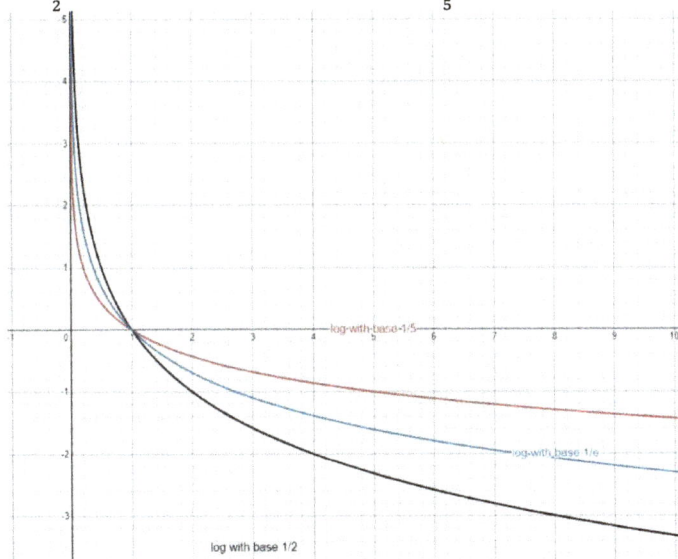

ii.   Transformations of the primary logarithmic functions

There are three types of transformations: horizontal translation, vertical translation, and scaling, which include positive and negative real numbers. When a scaling is negative, we will interpret the scaling as a reflection and a scaling.

$f(x) = a \cdot \log_b(x - h) + k$, b is a positive real number excluding 1. That is $0 < b < 1 \ or \ b > 1$.
Its domain is all real numbers larger than h, and its range is the real numbers.
The scaler $a \neq 0$; $h, k \in \mathbb{R}$; $\mathbb{R}$ is the set of real numbers. h is the horizontal translation, and k is the vertical translation.

Let $a = 2, b = 3 \ h = 5, k = -8$

The graph of $f(x) = 2\log_3(x - 5) - 8$ is in the green color.

The graph of $f(x) = \log_3 x$ is in red.

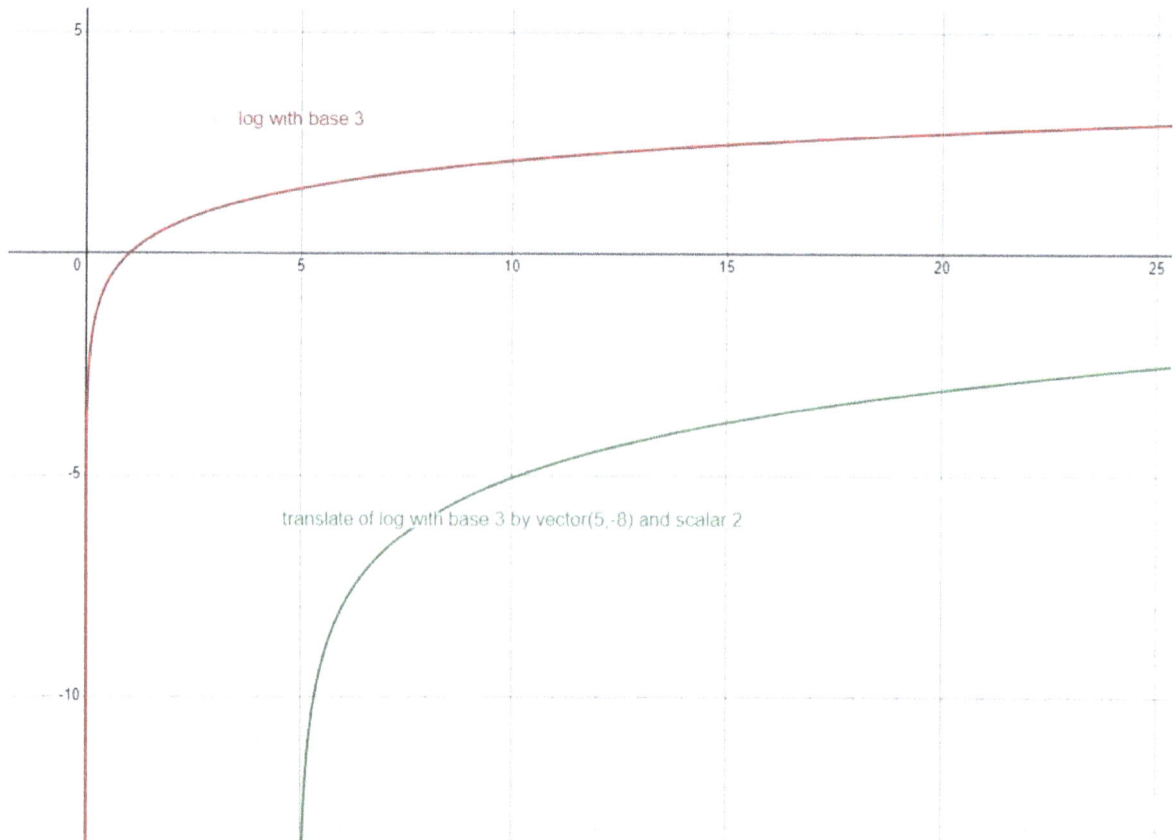

log with base 3

translate of log with base 3 by vector(5,-8) and scalar 2

iii.   Combine algebraic and logarithmic functions.

Five operations/combines on functions are +, -, *, /, °: addition, subtraction, multiplication, division, and composition.

Example 1: Present Value of an Annuity; Amortization

$$PV = PMT \cdot \frac{1 - (1+i)^{-n}}{i}$$

Where PV = present value of all payments (a debt)

     PMT = periodic payment

     i = rate per period

     n = number of periods

a. Express the number of periods to amortize a debt as a function with three variables PV, PMT, i.

b. Express the number of periods to amortize a debt as a function with one variable, the period interest rate. We fix PV and PMT values.

Solution 1. a

$$n(PV, PMT, i) = -\frac{\ln\left(1 - \frac{PV}{PMT} \cdot i\right)}{\ln(1+i)} \text{ with the condition } i < \frac{PMT}{PV}$$

The above number period function to retire debt is the result of an algebraic function of three variables and compositions of functions, the final division of two functions:

1. $f(x, y, z) = 1 - \frac{x}{y} \cdot z$, it is an algebraic function with three variables.
2. $g(x) = 1 + x$, it is a linear function with one variable.
3. $h(x) = \ln x$, it is the primary natural logarithmic Function.

$$n(PV, PMT, i) = -\frac{h(f(PV,PMT,i))}{h(g(i))}$$

Since the number of periods to amortize a debt function has the domain (0,0.12) and range (0,800), we will graph the function in logarithmic x-axe and y-axe. The graph is on the right.	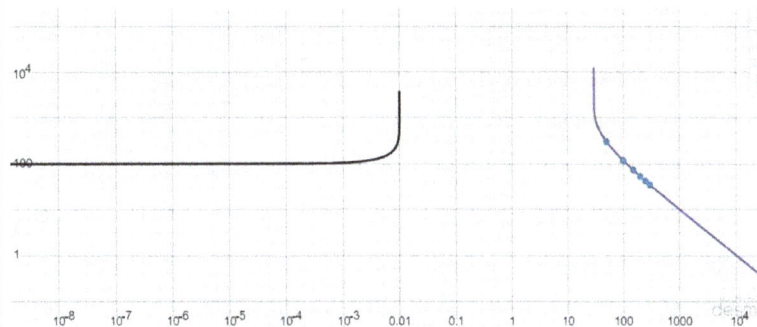

Solution 1. b

   b. Let the period interest rate be the independent variable, and fix the PV and PMT values.

a debt $PV = \$10000$, monthly payment $PMT = \$100$. The Function of the number of payments
is $n(x) = -\dfrac{\ln(1-100\ )}{\ln(1+x)}, i < 0.01$

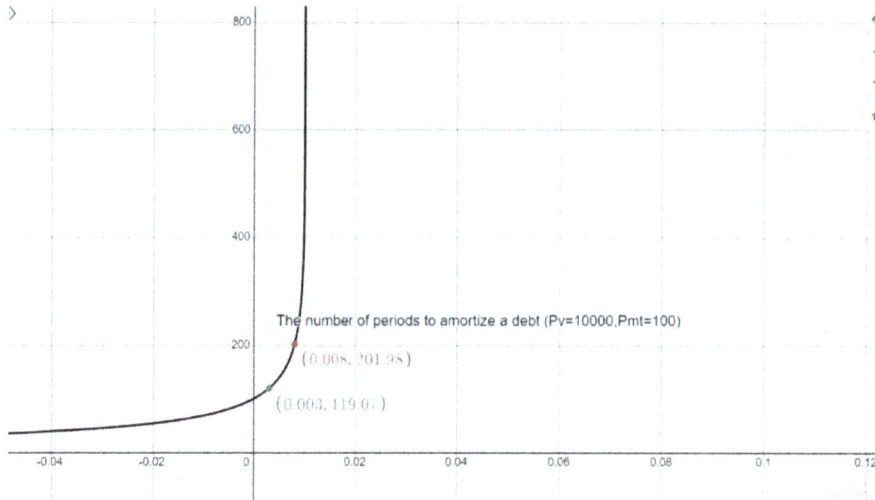

The number of periods to amortize a debt (Pv=10000,Pmt=100)

(0.008, 201.98)

(0.003, 119.07)

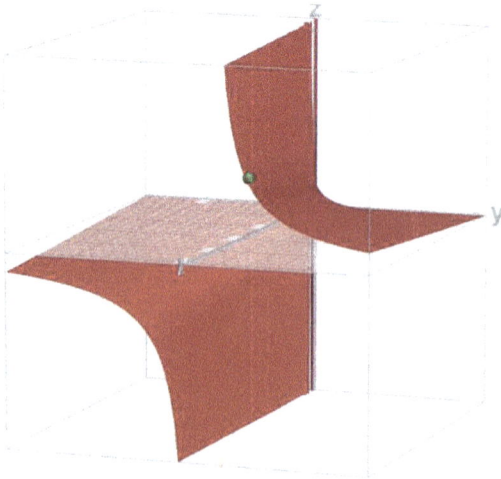

b. Let PV and PMT be the independent variables, the period interest rate be fixed.

a debt $PV = x$, monthly payment $PMT = y$. The monthly interest rate, $i = 0.005$.

The function of the number of payments is

$$n(x, y) = -\frac{\ln\left(1-\frac{0.005x}{y}\right)}{\ln(1+0.005)}, i < 0.01$$

The green point has the coordinates (10000,100,139). Which means to pay off a \$10000 debt if a person pays monthly of \$100, the person needs to pay 139 months (11 years and 7 months).

Chapter Three: Reasoning, Logic, and Sets

1. Sets
2. Basic Sets and Set Relations
3. Set Operations
4. Statements
5. Logic Statements
6. Standard Forms of Valid Arguments

Sets, logic, and reasoning are the bedrock upon which the entire edifice of human beings constructed mathematical systems. Sets, as defined collections of objects, provide the fundamental building blocks for mathematical structures and relationships. Through the rigorous application of logic, mathematicians formulate mathematical propositions, analyze them, and prove them with precision and clarity. This logical framework enables mathematicians to establish the validity of mathematical statements, leading to new theories and theorems. Furthermore, sound reasoning allows mathematicians to infer conclusions from established axioms and principles, fostering a deeper understanding of mathematical concepts and their interconnections. In essence, sets, logic, and reasoning form the indispensable foundation upon which mathematicians build the beauty and complexity of mathematical systems, guiding mathematicians in exploring the infinite realms of mathematical knowledge.

Set Examples

1. Natural Numbers or Counting Numbers
2. Whole Numbers
3. Integers
4. Rational Numbers
5. Irrational Numbers
6. Real Numbers
7. Complex Numbers
8. Polynomial Functions
9. Rational Functions
10. Algebraic Functions
11. Exponential Functions
12. Trigonometric Functions
13. Probability Distributions
14. Coordinate Systems

Basic Sets and Set Relations

1. Empty Set
2. Universal Set
3. Subsets
4. Sets

Set Operations or Form a New Set

1. Intersection Sets
2. Union Sets
3. Cartesian Product Sets
4. Complement of a Set

Statements

1. True or False Values
2. The Function maps from the set of statements to the set of {True, False}
3. A Simple Statement and Compound Statements
4. Statement Operations (Connectors) and Truth Value of a Compound Statement

Logical Statements and Logic

1. Negation
2. Conjunction
3. Disjunction
4. Conditional
5. Biconditional
6. Implication Statements
7. Equivalent Statements
8. Contradiction Statements
9. Contingency Statements
10. Tautology Statements
11. The Validity of an Argument Tool: Euler Diagrams

Arguments and Standard Forms of Valid Arguments

1. An Argument and a Valid Argument
2. Direct Reasoning
3. Contrapositive Reasoning
4. Transitive Reasoning
5. Disjunctive Reasoning

Chapter Four: Coordinates Systems and Vectors

I.   Cartesian Coordinates
1. 2D Graphs
   i. Transformations and Vectors
2. Conic in 2D and Rotations
3. Graphs of Trigonometric Functions in 2D
4. Trigonometric Functions in 3D
5. Quadric Surfaces
6. Gaussian Functions in 3D

II.   Parametric Coordinates
1. Parametric Equations and Graphs in 2D
2. Parametric Equations (Windows and Rockets)
3. Parametric Equations in 3D
4. Torus

III.   Polar Coordinates

IV.   Cylindrical Coordinates
1. Functions in Cylindrical Coordinates
2.

V.   Spherical Coordinates
1. Functions in Spherical Coordinates
2. Gaussian Functions in Spherical Coordinates
3. The number Six in Spherical Coordinates

VI.   Vectors
1. Vectors in 3D

I.   Cartesian Coordinates
  1. 2D Graphs in Cartesian Coordinates
          i.   Transformations and Vectors

Example: Use the graph of inverse sine and inverse cosine to generate a graph

Figure 1 is the graph of arccosine (inverse cosine):

Figure 1

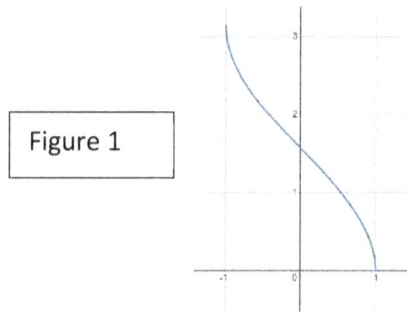

Figure 2

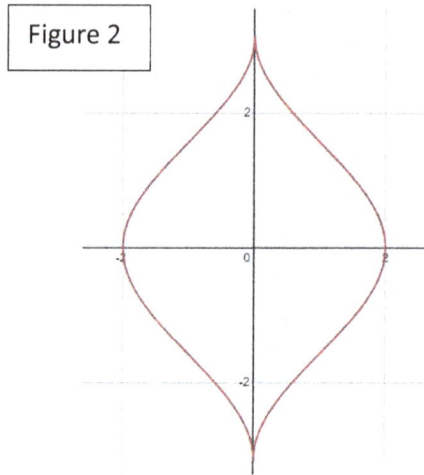

The Figure 2 the following four functions

3. F1 is Translation of arccosine with the vector (1,0)
4. F2 is the Reflection of F1 about the x-axe
5. F3 is the reflection of F1 about the y-axe
6. F4 is the reflection of F1 about the origin.

Figure 3

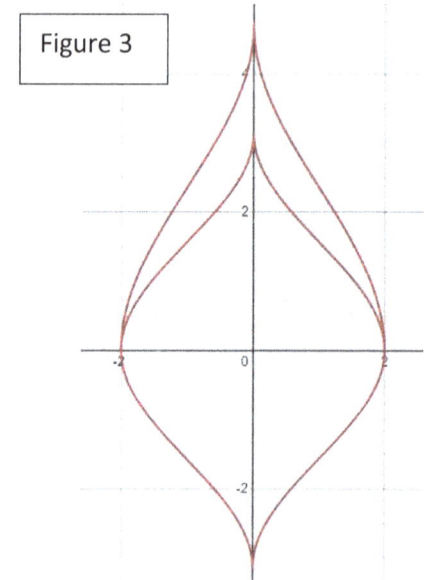

The Figure 3 adding two more functions

1. F5 is the vertical stretching of f1 by a factor of 3/2.
2. F6 is the reflection of F5 about the y-axe.

Figure 4 is the graph of arcsine (inverse sine):

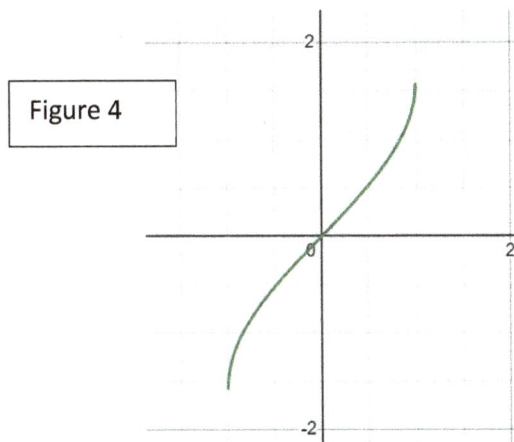

Figure 4

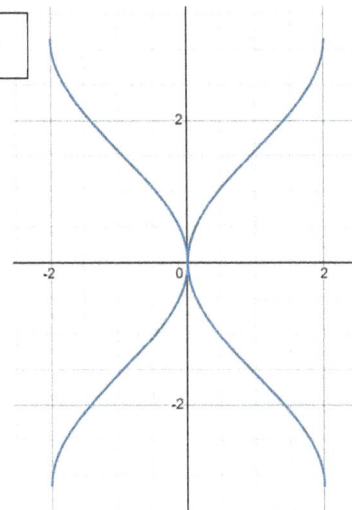

Figure 5

The Figure 5 the following four functions

7. F7 is translation of arcsine with the vector $(1, \frac{\pi}{2})$.
8. F8 is the Reflection of F7 about the x-axe
9. F9 is the reflection of F7 about the y-axe
10. F10 is the reflection of F7 about the origin.

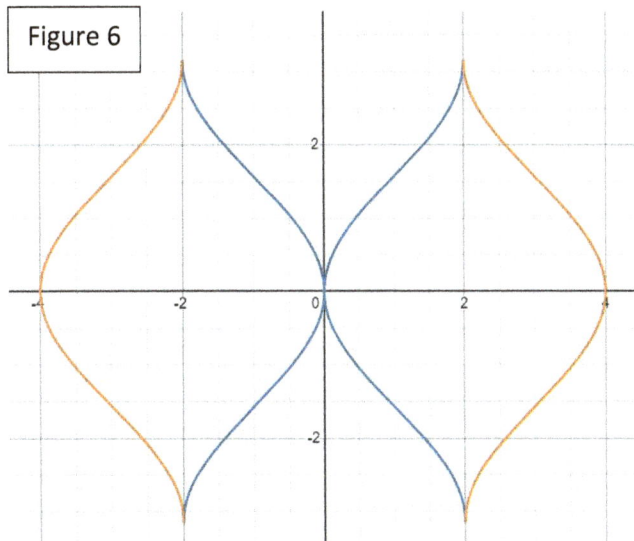

Figure 6

The Figure 6 the following four functions

1. F11 is the reflection of arcsine about the y-axe and translation with the vector $(3, \frac{\pi}{2})$

2. F12 is the reflection of F11 about the x-axe

3. F13 is the translation of arcsine with the vector $(-3, \frac{\pi}{2})$

4. F14 is the reflection of F13

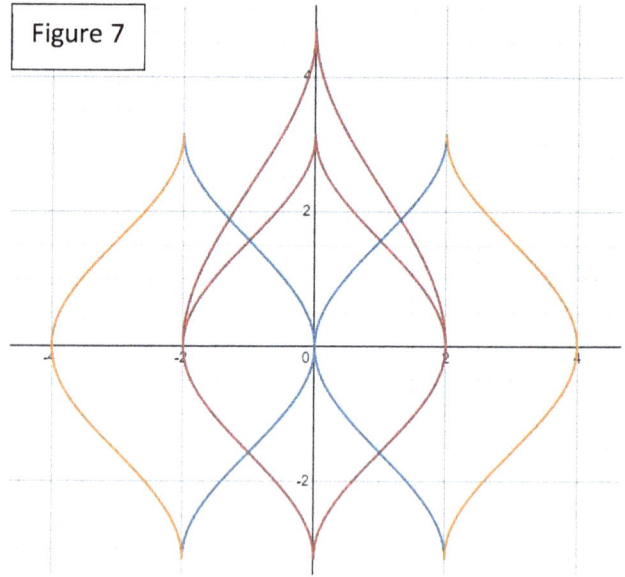

The Figure 7

Put all fourteen functions together.

Figure 7

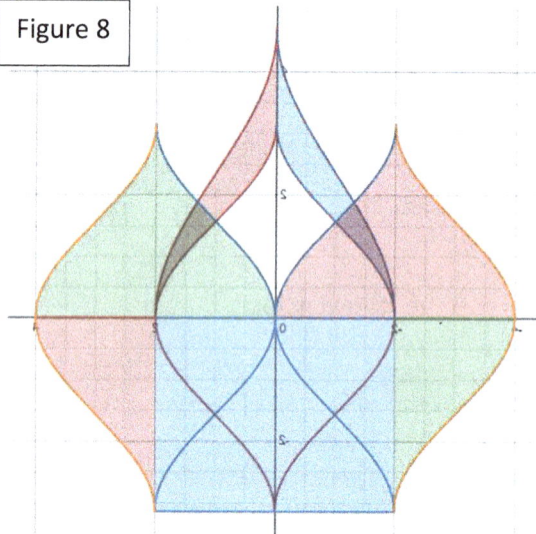

Figure 8

Using the inequalities relations among the above functions

Figure 8

2. Conic and Rotations

An equation in two variables is a relationship between the two variables. Let us look at some examples.

Examples

1. Circles
2. Parabolas
3. Ellipses
4. Hyperbolas
5. Other

Figure 9

Figure 9 on the right

The graph of circle, parabolas, ellipses, and hyperbola equations.

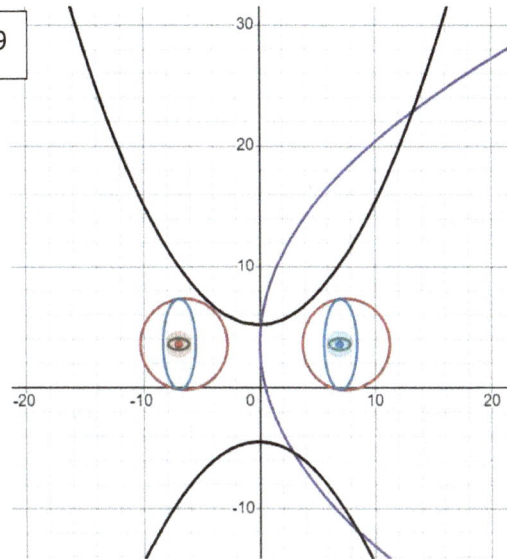

Figure 10 Rotate and transform the ellipse:

$$\frac{(x-2)^2}{2} + \frac{(y+1)^2}{5} = 1$$

This figure can be as a Venn diagram and a diagram for a quantum structure.

Figure 10

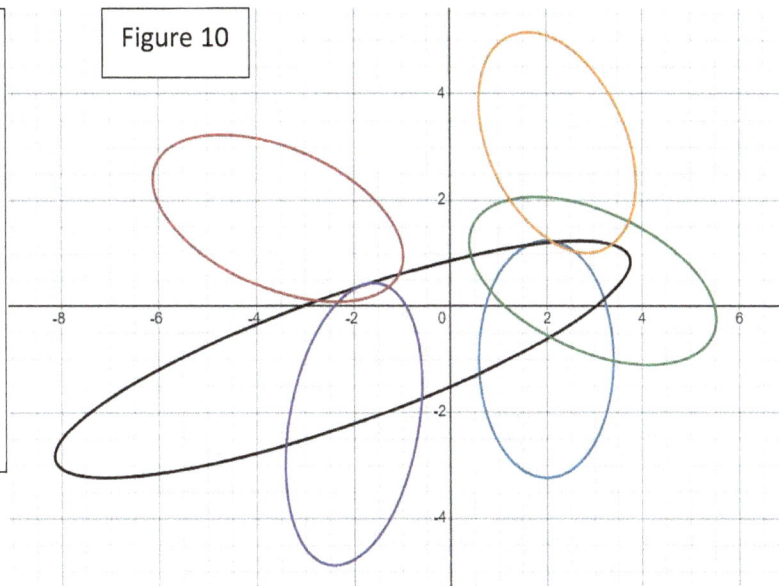

3. Graphs of Sine and Cosine Functions in 2D

Graph of Wave of sin(x) and sin(y).

$$\sin(y)\cos(x) = \sin(y^2)$$

$$-15 \leq x \leq 15, -20 \leq y \leq 20$$

Figure 11 below

The graph of an equation of sin(x) and sin(y).

Figure 11

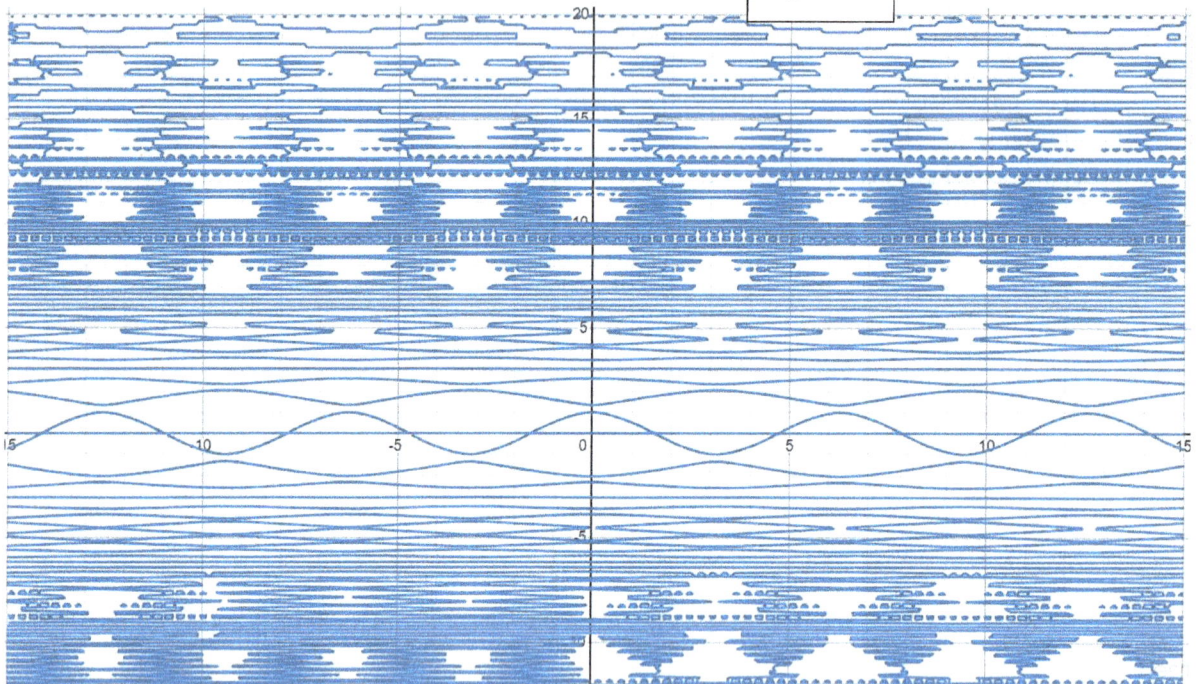

Graph of Wave of sin(x) and sin(y).

$$\sin(x)\cos(y) = \sin(x^2)$$

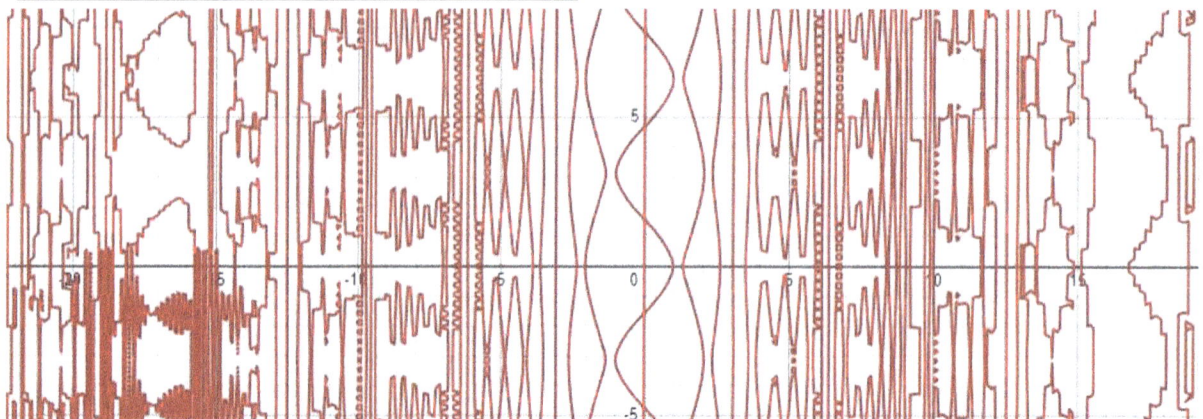

4. Trigonometry Functions in 3D

| Product of sine of x and sine of y $f(x,y) = \sin(x)\sin(y)$ | Sum of sine of x and sine of y $f(x,y) = \sin(x) + \sin(y)$ | Sine of the sum of x and y $f(x,y) = \sin(x+y)$ |

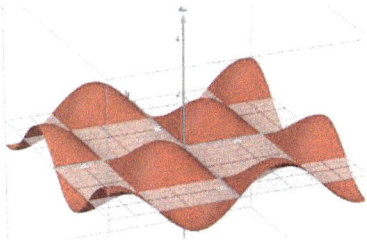 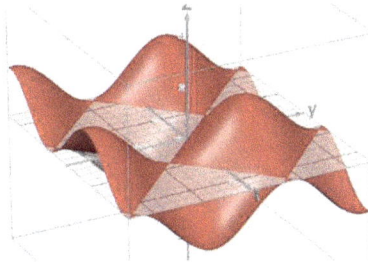 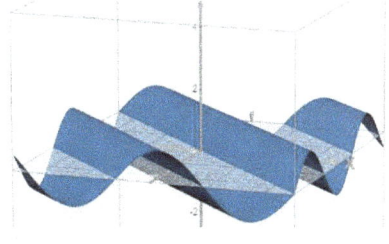

| Product of sine of x and cosine of y $f(x,y) = \sin(x)\cos(y)$ | Sum of sine of x and cosine of y $f(x,y) = \sin(x) + \cos(y)$ | Square of sin(x) and cos(y) $f(x,y) = (\sin(x) + \cos(y))^2$ |

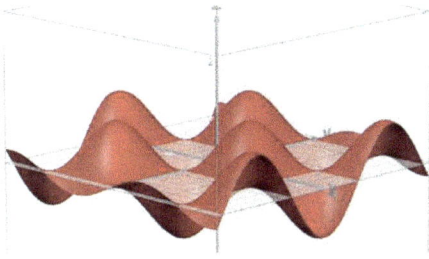 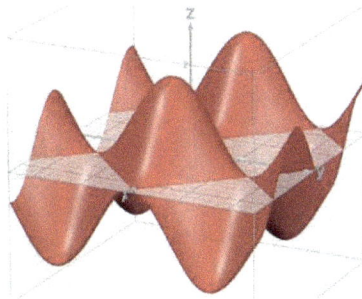 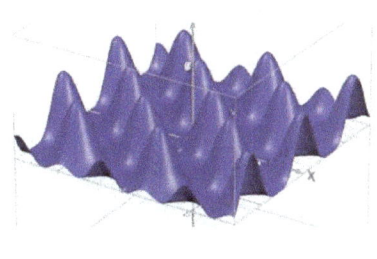

| Sum of tangent x and tangent y $f(x,y) = \tan(x) + \tan(y)$ | Product of tan x and tangent y $f(x,y) = \tan(x)\tan(y)$ | Tangent of sum x and y $f(x,y) = \tan(x+y)$ |

 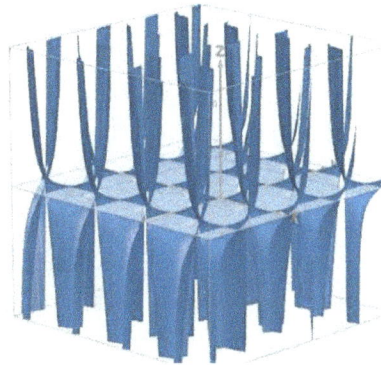 

Sum of tangent x and cot y	Product of tan x and cot y
$f(x,y) = \tan(x) + \cot(y)$	$f(x,y) = \tan(x)\cot(y)$

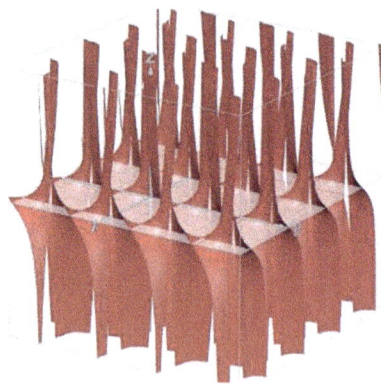

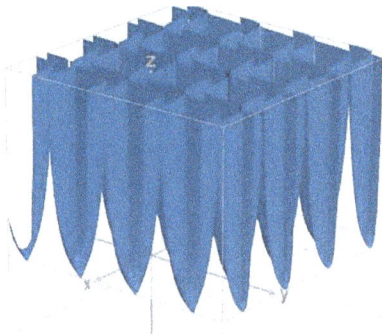

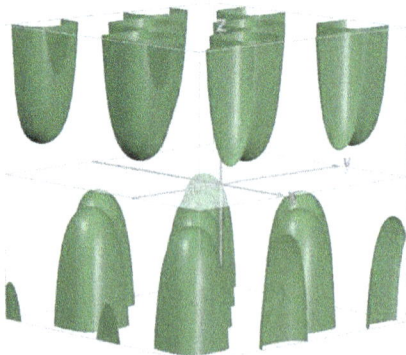

Square of sum tan x and cot y	Sum of cosecant x and sec y
$f(x,y) = (\tan(x) + \cot(y))^2$	$f(x,y) = \csc(x) + \sec(y)$

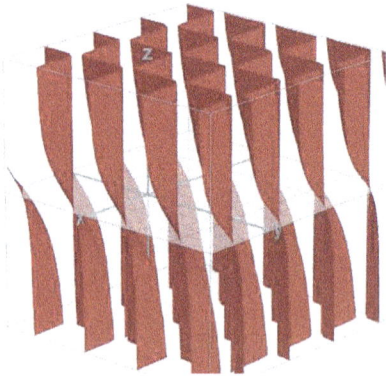

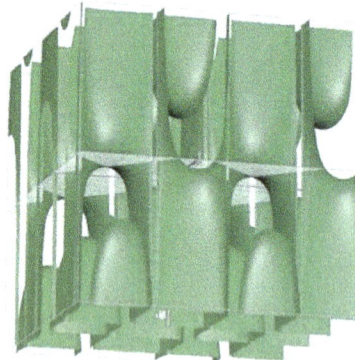

Product of cosecant x and sec y	Square of sum csc x and sec y
$f(x,y) = \csc(x)\sec(y)$	$f(x,y) = (\csc(x) + \sec(y))^2$

3D Cartesian Coordinates

Examples

1. Quadric Surfaces
2. Gaussian Functions

5. Quadric Surfaces (3D Cartesian Coordinates)

Examples Quadric Surfaces

1. Sphere
2. Ellipsoid
3. Elliptic Paraboloid
4. Right Circular Cone
5. Double Cone
6. Elliptic Cone
7. Right Circular Cylinder
8. Right Ellipse Cylinder
9. Right Parabolic Cylinder
10. Hyperboloid of One Sheet
11. Hyperboloid of Two Sheet
12. Hyperbolic Paraboloid

Equations for the graphs
1. $(x-1)^2 + (y-1)^2 + (z-1)^2 = 1$
2. $0.5x^2 + 2y^2 + z^2 = 1$
3. $f(x,y) = 0.5x^2 + 2y^2$
4. $f(x,y) = \sqrt{x^2 + y^2}$
5. $x^2 + y^2 = z^2$
6. $f(x,y) = \sqrt{x^2 + 3y^2}$
7. $x^2 + y^2 = 1$
8. $0.5x^2 + 2y^2 = 1$
9. $f(x,y) = x^2 - 1$
10. $-x^2 + 2y^2 + z^2 = 1$
11. $x^2 + 2y^2 - z^2 = -1$
12. $f(x,y) = 0.5x^2 - 2y^2$

Sphere

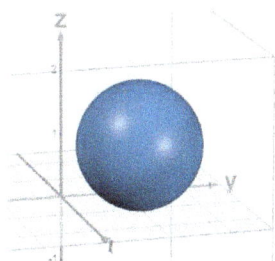

Ellipsoid

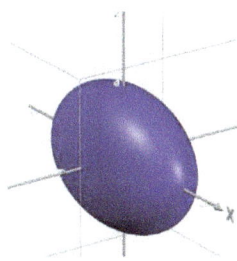

Elliptic Paraboloid

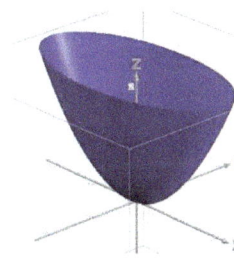

Right Cylinder Cone

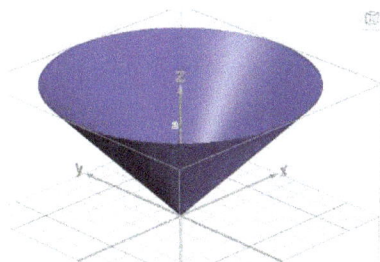

Right Double Cones

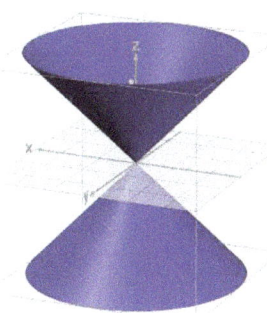

Elliptic Cone

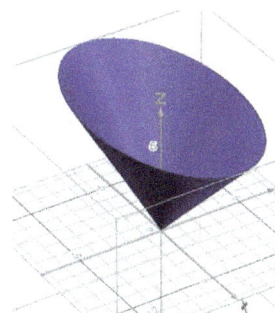

Right Circular Cylinder

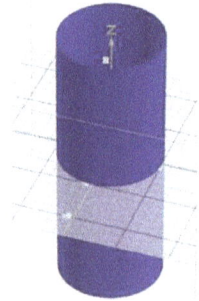

Right Ellipse Cylinder

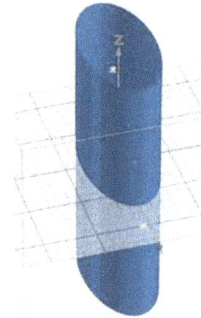

Right Parabolic Cylinder

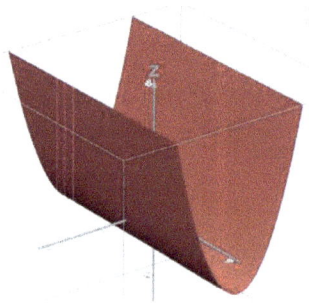

Hyperboloid of One Sheet

Hyperboloid of Two-Sheet

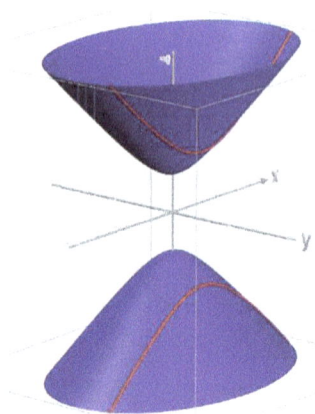

Hyperbolic Paraboloid

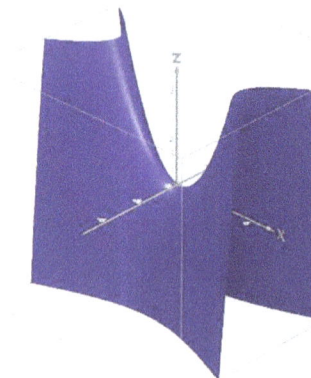

6. Gaussian Functions in 3D

$$f(x, y) = \frac{1}{y\sqrt{2\pi}} e^{-\frac{x^2}{2y^2}}$$

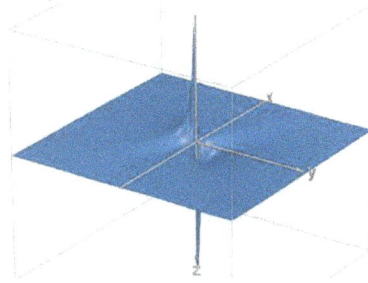

$$f(x, 1) = \frac{1}{\sqrt{2\pi}} e^{-\frac{x^2}{2}}$$

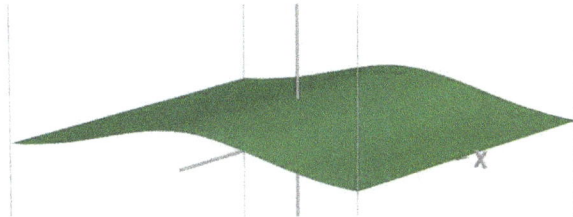

$$f(\sin(x), y) = \frac{1}{y\sqrt{2\pi}} e^{-\frac{(\sin(x))^2}{2y^2}}$$

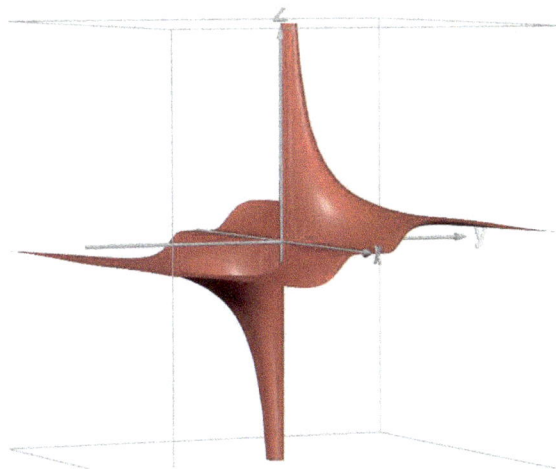

$$f(x, \sin(y)) = \frac{1}{\sin(y)\sqrt{2\pi}} e^{-\frac{x^2}{2(\sin(y))^2}}$$

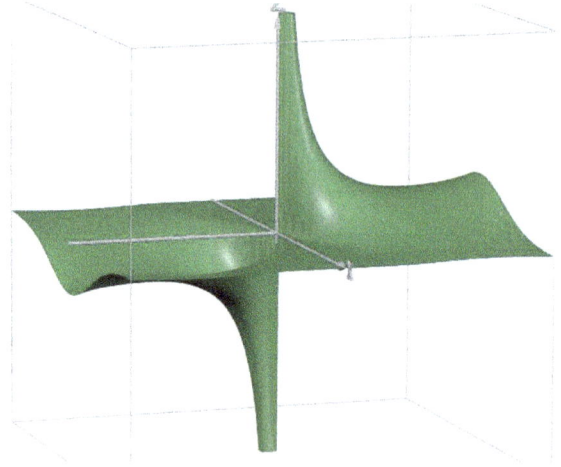

$$f(\sin(x), \sin(y)) = \frac{1}{\sin(y)\sqrt{2\pi}} e^{-\frac{(\sin(x))^2}{2(\sin(y))^2}}$$

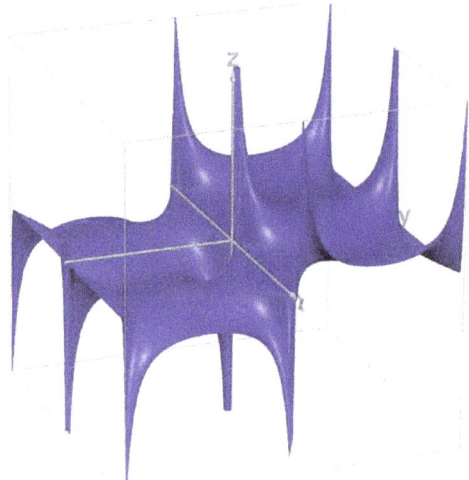

$$f(\sin(2x), \sin(y)) = \frac{1}{\sin(y)\sqrt{2\pi}} e^{-\frac{(\sin(2x))^2}{2(\sin(y))^2}}$$

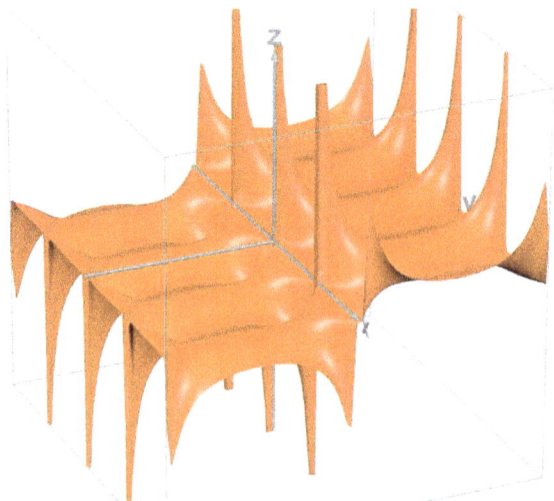

I. Parametric Coordinates
   1. Parametric Equations and Graphs in 2D
   2. Parametric Equations (Windows and Rockets)
   3. Parametric Equations in 3D (Lamp)
   4. Torus and Modifications

II. Parametric Coordinates

1. Parametric Equations and Graphs in 2D

$$\left(\mp(\sin(2t)+2), \cos^2\left(t-\frac{\pi}{4}\right)\right) \text{ green}$$
$$0 \leq t \leq 2\pi$$
$$(t, -\sqrt{1-t^2}) \quad -1 \leq t \leq 1 \text{ purple}$$
$$(\tfrac{1}{2}t, \sqrt{2-t^2}) \quad -\sqrt{2} \leq t \leq \sqrt{2} \text{ purple}$$
$$(\sin(2t), t) \quad 0 \leq t \leq 2\pi \text{ red}$$
$$(t-\pi, \cos t - 1) \quad -1 \leq t \leq 1 \text{ blue}$$
$$(t, 0) \quad -1 \leq t \leq 1 \text{ Black}$$

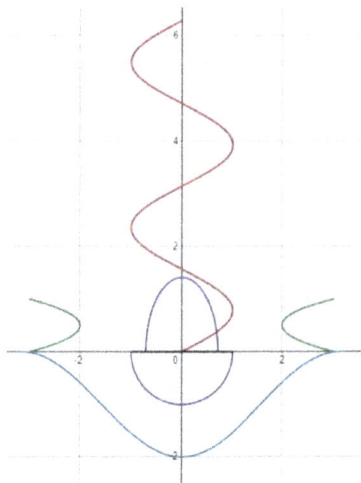

$$(\sin(2t), \sin(3t))$$
$$0 \leq t \leq 2\pi$$

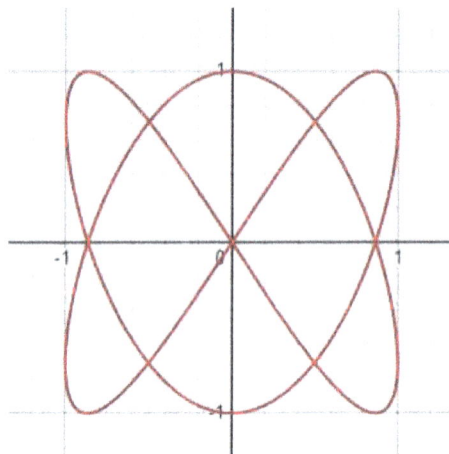

$$f_1(a,b) = ab - \sin(ab)$$
$$f_2(a,b) = 1 - \cos(ab)$$
$$(f_1(1,t), \mp f_2(2,t)) \text{ purple}$$
$$(\mp f_2(1,t), \mp f_1(2,t)) \text{ Red}$$
$$(\mp f_2(2,t), f_1(2,t)) \text{ Green}$$
$$-10 \leq t \leq 10$$

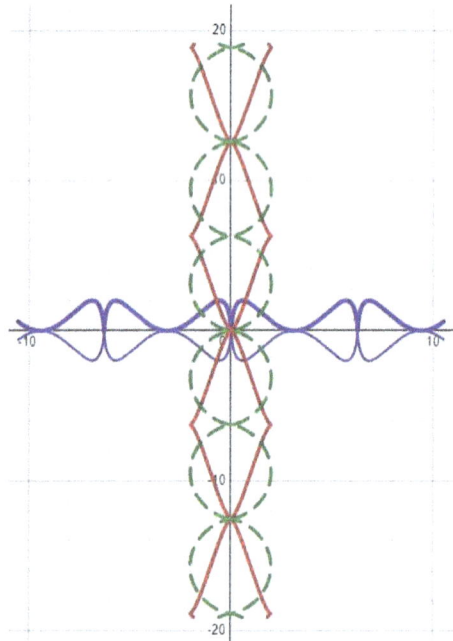

2. Parametric Equations (Windows and Rockets)

Figure 12 below Parametric Equations: Windows and Rockets

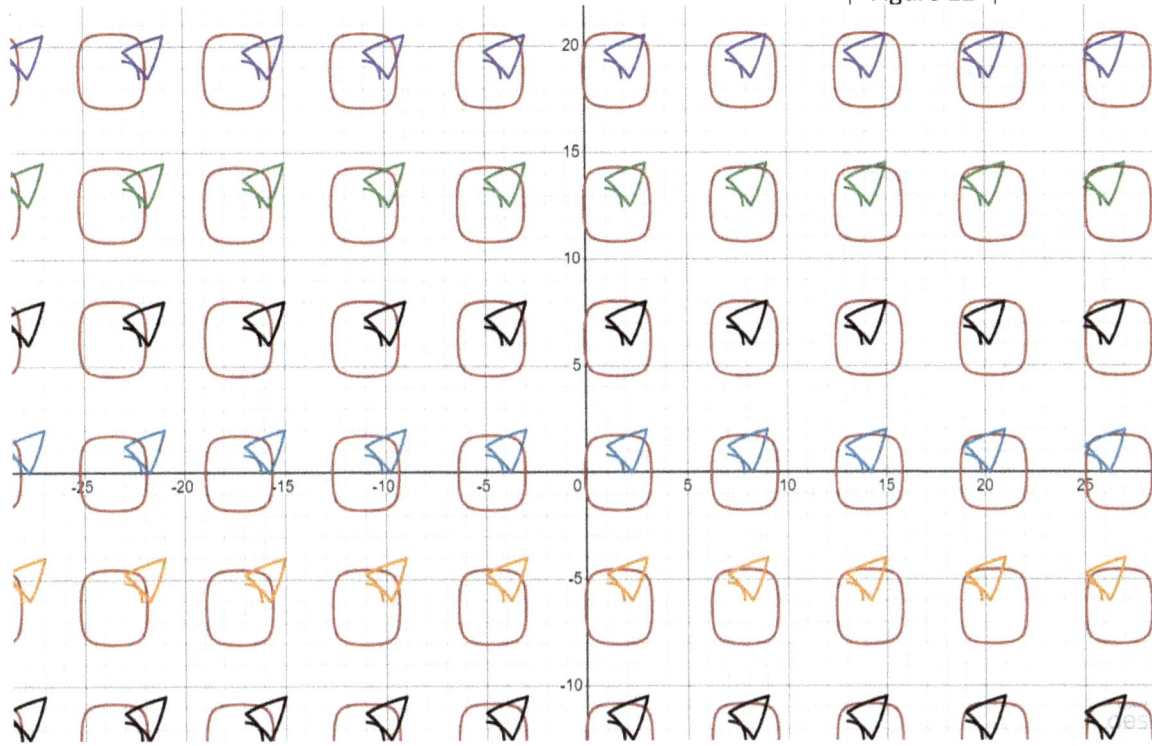

Figure 12

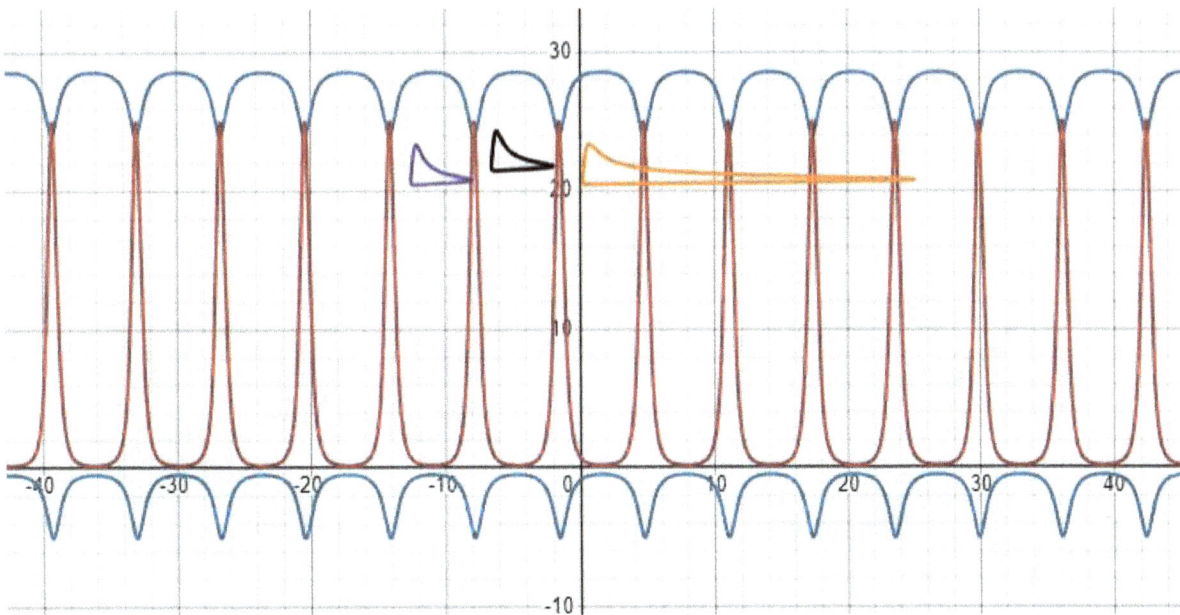

3. Parametric Equations and Graphs in 3D

Example Lamp

Three curves in 3D as the following

$$g_0(x) = \frac{(\pi x)^{\frac{1}{4}} + \sin\left(\frac{\pi}{2}x\right) + \cos(\pi x)}{\pi(x+1)} \{x \leq 7\} \text{ \{the wire\}}$$

$$g_1(x) = \frac{(\pi x)^{\frac{1}{4}} + \sin\left(\frac{\pi}{2}x\right) + \cos(\pi x)}{\pi(x+1)} + 0.5 \text{ \{the curve for the base of the lamp\}}$$

$$g_2(x) = \frac{(\pi(x-6))^{\frac{1}{4}} + \sin\left(\frac{\pi}{2}(x-6)\right) + co\ (\pi(x-6))}{\pi(x-5)} + 1.7 \text{ \{the curve for the cover of the lamp\}}$$

Two Surfaces by the rotating the $g_1(x)$ and $g_2(x)$:

$$(u, g1(u)\cos(v), g_1(u)\sin(v))$$
$$0 \leq u \leq 7 \qquad \text{\{The base of the lamp\}}$$
$$0 \leq v \leq 10$$

$$(u, g1(u)\cos(v), g_1(u)\sin(v))$$
$$0 \leq u \leq 7 \qquad \text{\{The cover of the lamp\}}$$
$$0 \leq v \leq 10$$

The Equation for bulb

$$(x - 7.5)^2 + y^2 + z^2 = 1$$

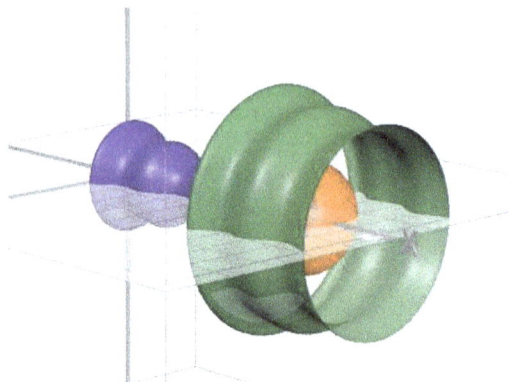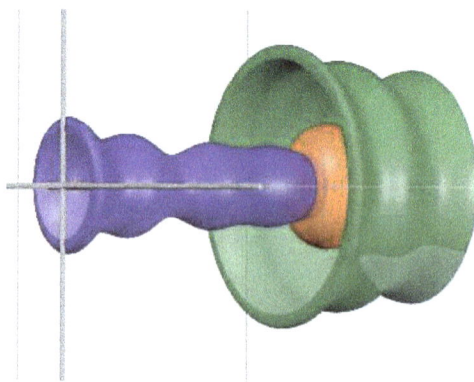

4. Torus and Modifications (Parametrization)

Torus:
$f_1(u,v) = (4 + \cos(u))\cos(v)$
$f_2(u,v) = (4 + \cos(u))\sin(v)$
$f_3(u) = \sin(u)$
Parametrized Coordinates in 3D
$(f_1(u,v), f_2(u,v), f_3(u,v))$
$\qquad 0 \le u, v \le 2\pi$

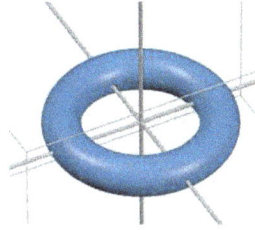

$(\dfrac{f_1(u,v)}{2+\cot(u)}, \dfrac{f_2(u,v)}{2+\cos(v)}, 3(f_3(u) + 2 + \cos(u+v)))$

$(\dfrac{f_1(u,v)}{2+\sin(u)}, f_2(u,v), f_3(u))$

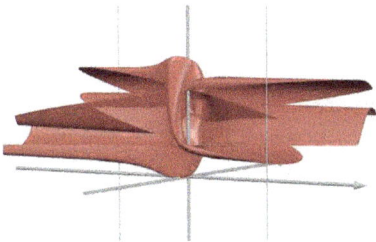

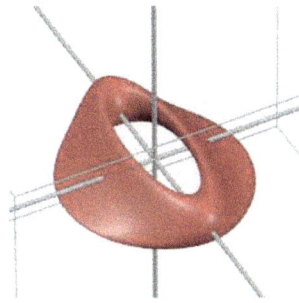

$(\dfrac{f_1(u,v)}{2+\sin(u)}, \dfrac{f_2(u,v)}{2+\cos(v)}, f_3(u))$

$(\dfrac{f_1(u,v)}{2+\sin(u)}, \dfrac{f_2(u,v)}{2+\cos(v)}, \dfrac{f_3(u)}{2-\cos(u+v)})$

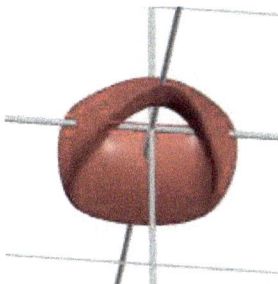

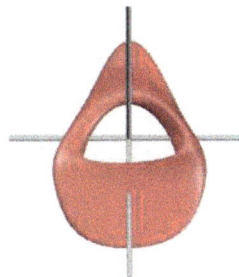

$$\left(\frac{f_1(u,v)}{2+\sin(u)}, \frac{f_2(u,v)}{2+\cos(v)}, \frac{f_3(u)}{2+\cos(u+v)}\right)$$

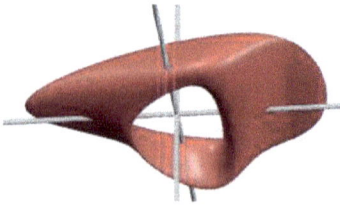

$$\left(\frac{f_1(u,v)}{2+\sin(u)}, \frac{f_2(u,v)}{2+\cos(v)}, f_3(u)+2\right.$$

A Mobius
Strip

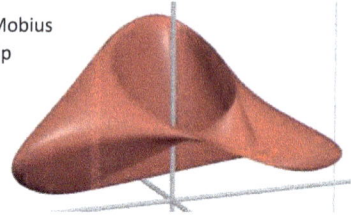

$$\left(\frac{f_1(u,v)}{2+\tan(u)}, \frac{f_2(u,v)}{2+\cos(v)}, f_3(u)+2+\cos(u+v)\right.$$

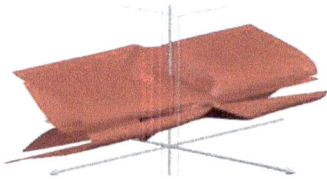

$$\left(\frac{f_1(u,v)}{2+\cot(u)}, \frac{f_2(u,v)}{2+\cos(v)}, f_3(u)+2+\cos(u+v)\right.$$

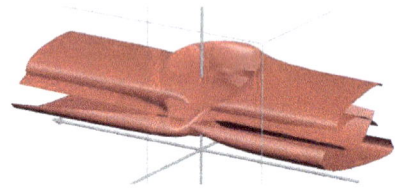

$$\left(\frac{f_1(u,v)}{2+\cot(u)}, \frac{f_2(u,v)}{2+\cos(v)}, 2f_3(u)+2+\cos(u+v)\right.$$

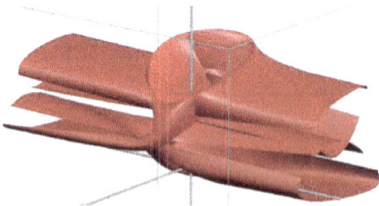

$$\left(\frac{f_1(u,v)}{2+\csc(u)}, \frac{f_2(u,v)}{2+\cos(v)}, f_3(u)+2+\cos(u+v)\right.$$

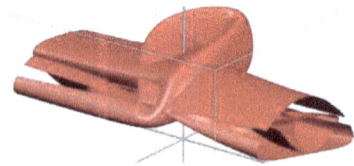

$$(f_1(u,v)\sin(v), f_2(u,v)(\cos(v)+\sin(u), f_3(u)\cos(v))$$

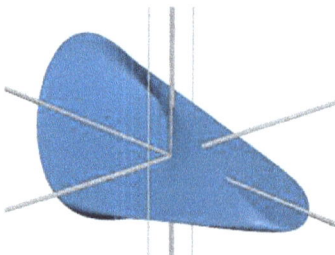

$$(f_1(u,v), f_2(u,v)(\cos(v)+\sin(u)), f_3(u))$$

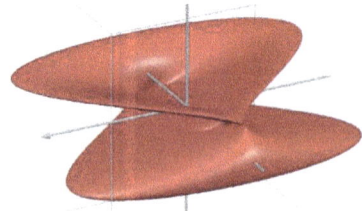

78

$$\left(\frac{f_1(u,v)}{2+\csc(u)}, \frac{f_2(u,v)}{2+\cos(v)}, f_3(u)+2v+\cos(u\right.$$

$$(f_1(u,v), f_2(u,v)\sin(u), f_3(u)\cos(v))$$

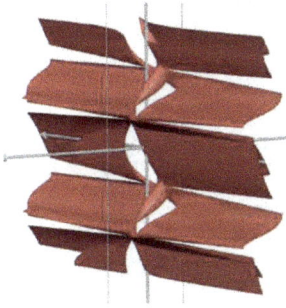

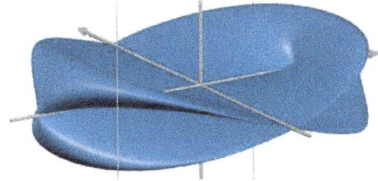

$$(f_1(u,v), f_2(u,v)\sin(u), f_3(u)v))$$

$$(f_1(u,v), f_2(u,v)\sin(u+v), f_3(u))$$

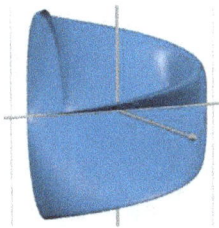

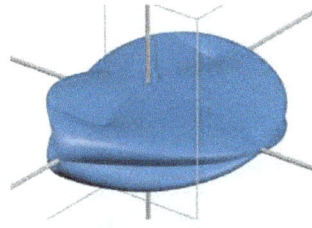

$$(f_1(u,v), f_2(u,v)\sin(u), f_3(u)$$

$$(f_1(u,v)\sin(u), f_2(u,v)(\cos(v)+\sin(u), f_3(u)\cos(v))$$

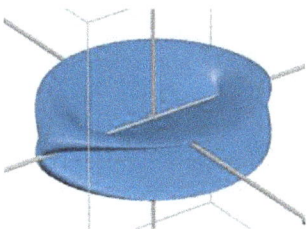

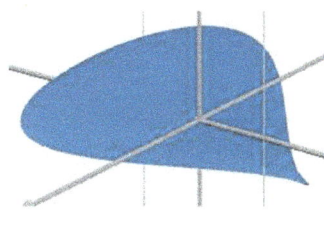

$$(\sin(f_1(u,v)), f_2(u,v)(\cos(v)+\sin(u), \sin(f_2(u)))$$

$$(\sin(f_1(u,v)), f_2(u,v)(\cos(v)+\sin(u), f_3(u))$$

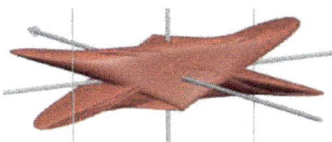

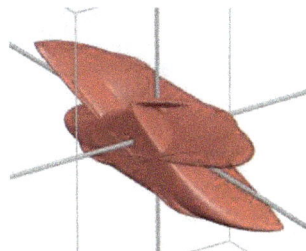

$(\sin(f_1(u,v)), f_2(u,v), \sin(f_3(u)))$

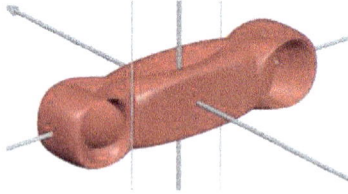

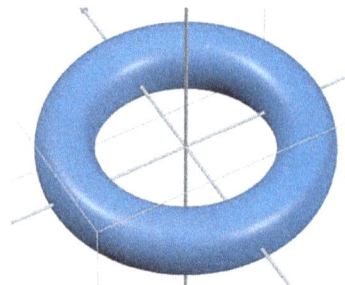

$(\sin(f_1(u,v)), \cos(f_2(u,v)), \sin(f_3(u)))$

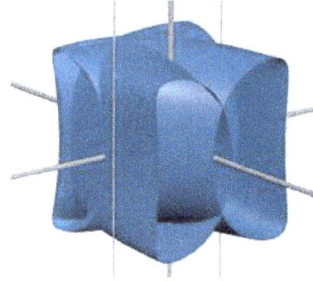

$(\sin(f_1(u,v)), \sin(f_2(u,v)), \sin(f_3(u)))$

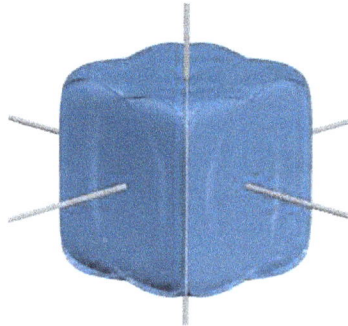

$(\sin(f_1(u,v)), \sin(f_2(u,v)), \sin(v + f_3(u)))$

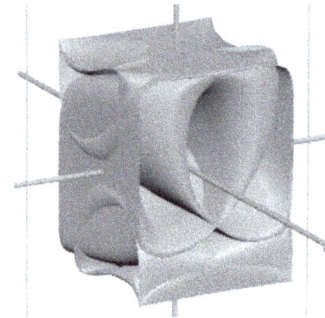

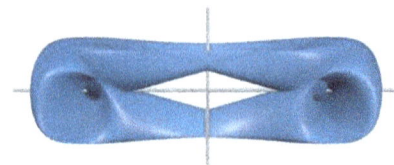

$(f_1(u,v), \sin(f_2(u,v)), \sin(v + f_3(u)))$

$(f_1(u,v)), f_2(u,v), \sin(f_3(u)))$

$(f_1(u,v), f_2(u,v), \sin(f_3(u)) + v)$

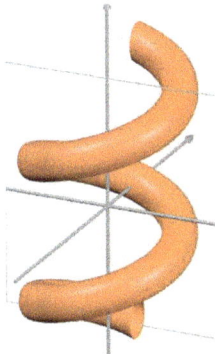

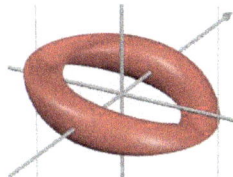

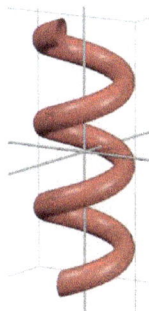

$(f_1(u,v)), f_2(u,v)), \sin(f_3(u) + v))$

$(f_1(u,v), f_2(u,v), f_3(u) + v)$

$(f_1(u,v), f_2(u,v) + \sin(v), f_3(u))$

$(f_1(u,v) + \sin(u), f_2(u,v) + \sin(v), f_3(u))$

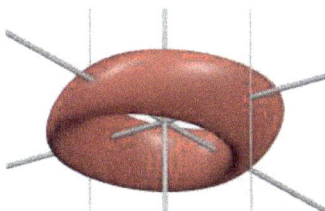

$(f_1(u,v) + \sin(u), f_2(u,v) + \sin(v), f_3(u) + \cos(u))$

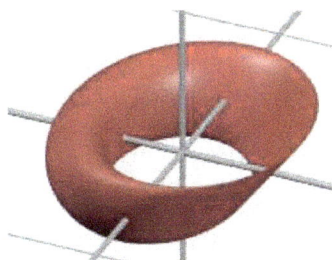

$(f_1(u,v) + \sin(u), f_2(u,v) + \sin(v), f_3(u) + \sin(u))$

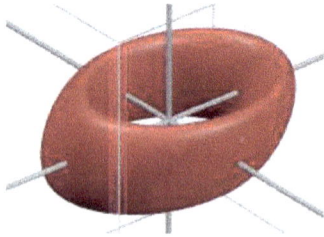

$(f_1(u,v) + \sin(u), f_2(u,v) + \sin(v), f_3(u)$
$+ \sin(9v) + u$
$\{0 \leq u \leq 2\pi\}$

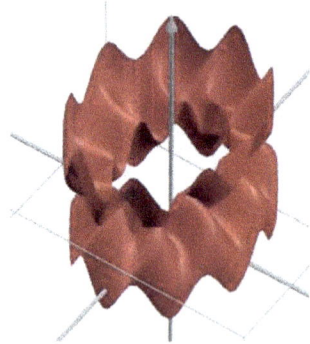

$(f_1(u,v) + \sin(u), f_2(u,v) + \sin(v),$
$f_3(u) + \sin(v+u) \{0 \leq u,v \leq 2\pi\}$

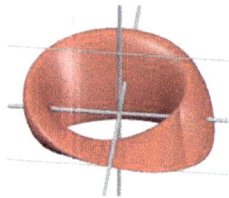

$(f_1(u,v) + \sin(u+v), f_2(u,v) + \sin(v),$
$f_3(u) + \sin(u+v) \{0 \leq u,v \leq 2\pi\}$

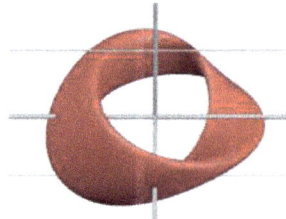

$(f_1(u,v) + \sin(u+v), f_2(u,v) + \sin(v),$
$f_3(u) + \sin(2(u+v)) \{0 \leq u,v \leq 2\pi\}$

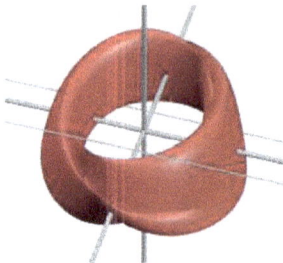

$(f_1(u,v) + \sin(u+v), f_2(u,v) + \sin(v),$
$f_3(u) + \sin(0.5(u+v)) \{0 \leq u,v \leq 2\pi\}$

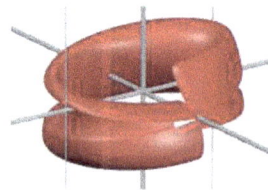

$(f_1(u,v) + \sin(u), f_2(u,v) + \sin(v),$
$f_3(u) + \sin(9v) + u \{-2\pi \leq u \leq 2\pi\}$

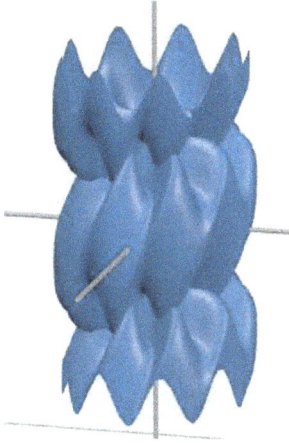

$(f_1(u,v) + u + v, f_2(u,v) + \sin(v), f_3(u) +$
$\sin(u+v)) \{-2\pi \leq u,v \leq 2\pi\}$

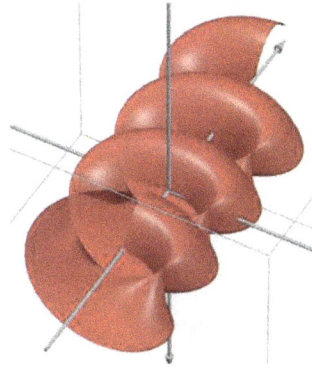

$(f_1(u,v) \sin(u+v), f_2(u,v) + \sin(v), f_3(u) +$
$\sin(u+v) \{0 \leq u,v \leq 2\pi\}$

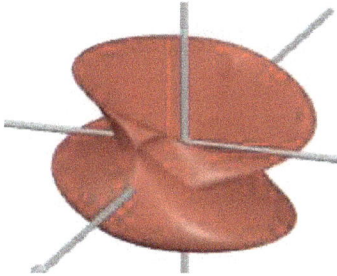

$(f_1(u,v) \sin(2(u+v)), f_2(u,v) + \sin(v),$
$f_3(u) + \sin(u+v)) \{0 \leq u,v \leq 2\pi\}$

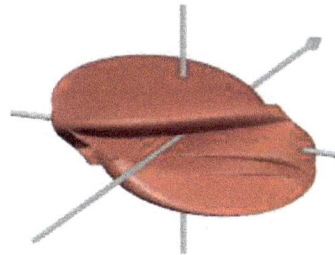

$(f_1(u,v) + u + v, f_2(u,v) + u + \sin(v),$
$f_3(u) + \sin(u+v)) \{-2\pi \leq u,v \leq 2\pi\}$

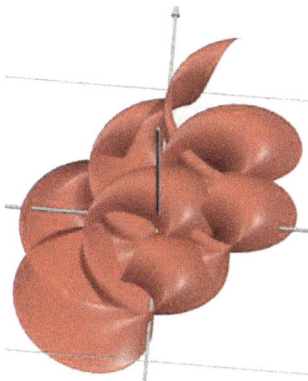

$(f_1(u,v) + u + v, f_2(u,v) + u + v +$
$\sin(v), f_3(u) + \sin(u+v))$
$\{-2\pi \leq u,v \leq 2\pi\}$

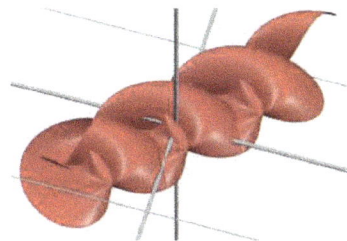

83

$$(f_1(u,v) + u + v, f_2(u,v) + u + \sin(v),\ f_3(u) + v + \sin(u + v))$$
$$\{-2\pi \le u, v \le 2\pi\}$$

III.   Polar Coordinates (2 pages)
    1. Roses
            i.   n is odd, n-leaved
            ii. n is even, 2n-leaved
    2. Cardioids and Lemniscates
    3. Graphs of Systems Inequalities
    4. Satellite Orbits

III.   Polar Coordinates
        3.   Roses

$r = a \cdot \cos(n\theta)$ or $r = a \cdot \sin(n\theta)$

n is odd; the n-leaved roses

Examples $n = 1, 3, 2, 4$

$r = 2\cos(\theta)$ in red

$r = 2\sin(\theta)$ in purple

$0 \le \theta \le \pi$

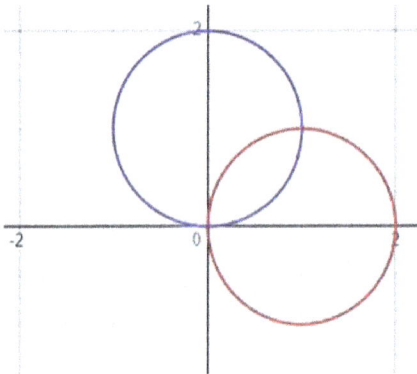

$r = 2\cos(3\theta)$ in red

$r = 2\sin(3\theta)$ in purple

$0 \le \theta \le \pi$

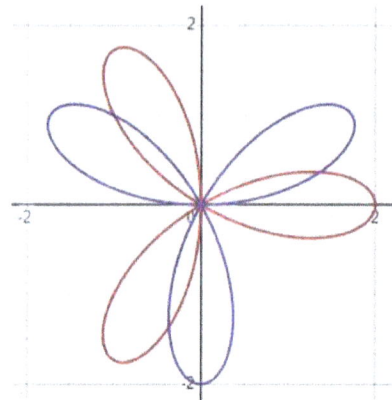

$r = 2\cos(2\theta)$ in red

$r = 2\sin(2\theta)$ in purple

$0 \le \theta \le \pi$

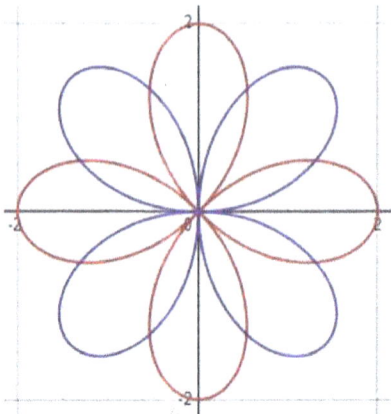

$r = 2\cos(4\theta)$ in red

$r = 2\sin(4\theta)$ in purple

$0 \le \theta \le \pi$

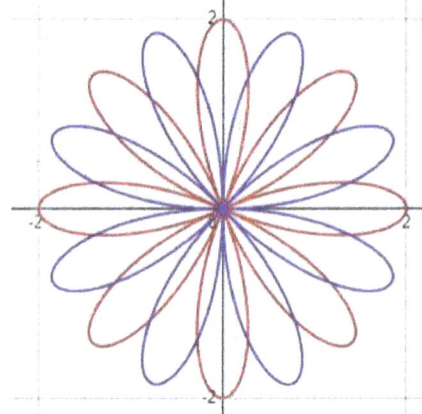

2. Cardioids; Lemniscates; Balloons; Limacons

$r = a(1 \mp \cos(n\theta))$ or $r = a(1 \mp \sin(n\theta))$

When n=1, the curve is a cardioid. When n=2, the curve is a Lemniscate.

When n > 2, the curve is n-balloons

Examples $n = 1, 2, 3$

$r = 2(1 - \cos(\theta))$ in red

$r = 2(1 - \sin(\theta))$ in purple

$$0 \leq \theta \leq \pi$$

$r = 2(1 + \cos(2\theta))$ in red

$r = 2(1 + \sin(2\theta))$ in purple

$$0 \leq \theta \leq \pi$$

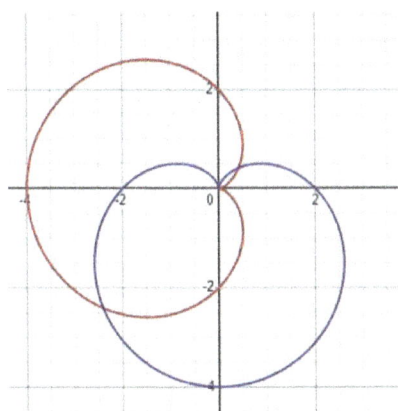

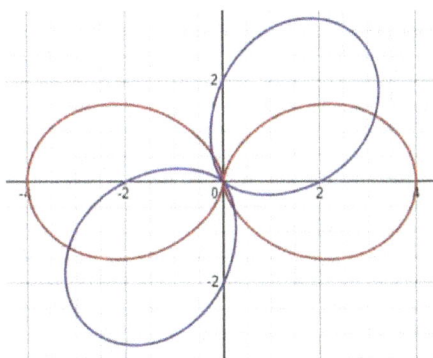

$r = 2(1 + \cos(3\theta))$ in red

$r = 2(1 + \sin(3\theta))$ in blue

$$0 \leq \theta \leq \pi$$

$r = 2(1 + 2\cos(3\theta))$ in red

$r = 2(1 + 2\sin(3\theta))$ in purple

$$0 \leq \theta \leq \pi$$

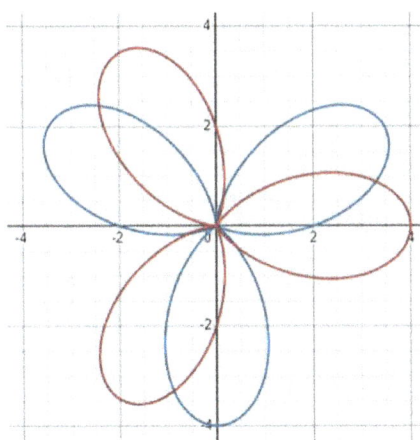

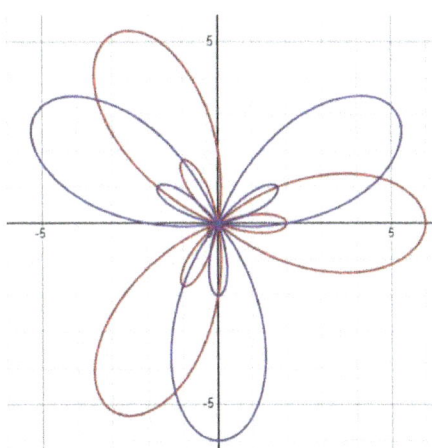

3. Graphs of systems of inequalities

$$r \leq 2(1 + \cos(2\theta))$$
$$r \leq 2(1 + \sin(\theta))$$
$$r \leq 2(1 + \sin(3\theta))$$

$$r \leq 2(1 + \cos(2\theta))$$
$$r \geq 2(1 + \sin(\theta))$$
$$r = 2(1 + \sin(3\theta))$$

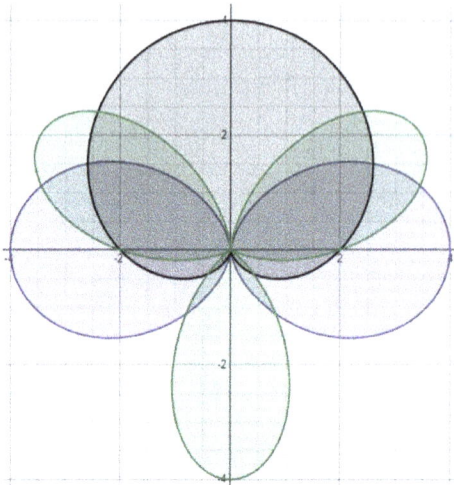

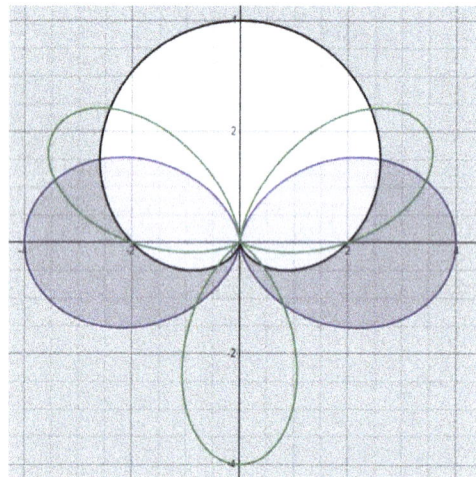

$r \leq 1$ red

$r \leq \frac{2(1 + c \quad (2\theta))}{2(1 + si \quad (2\theta))}$ blue

$r \leq \frac{2(1 + \cos(2\theta))}{2(1 + \sin(\theta))}$ purple

Roses

$$r \geq 3\sin(4\theta)$$
$$r \leq 2(1 + \sin(5\theta))$$

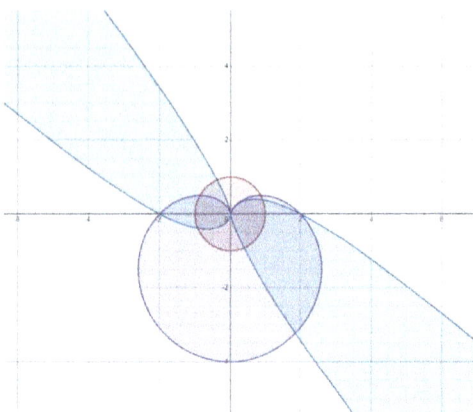

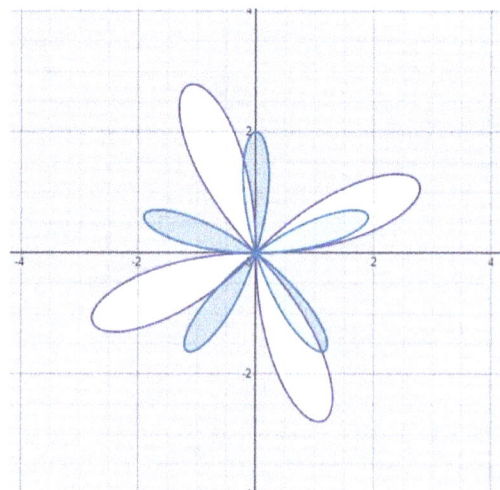

4. Satellite Orbits in the Solar System

$S_1(a, b, \theta) = \frac{a(1-b^2)}{1+b \cdot \cos(\theta)}$, $L_1 and L_2$ are lists contains a and b values

a is the average distance in astronomical units from the sun.

b is a eccentricity.

$r = S_1(L_1[i], L_2[i], \theta)$ for $i = 1, 2, \ldots, 9$

Satellite Orbits in the solar system

Satellite	L1 = a	L2 = b
Mercury	0.39	0.206
Venus	0.78	0.007
Earth	1.00	0.017
Mars	1.52	0.093
Jupiter	5.20	0.048
Saturn	9.54	0.056
Uranus	19.20	0.047
Neptune	30.10	0.009
Pluto	39.40	0.249

Source: Zeilik, M. S. Gregory, and E. Smith. Introductory Astronomy and Astrophysics, Saunders College Publishers.

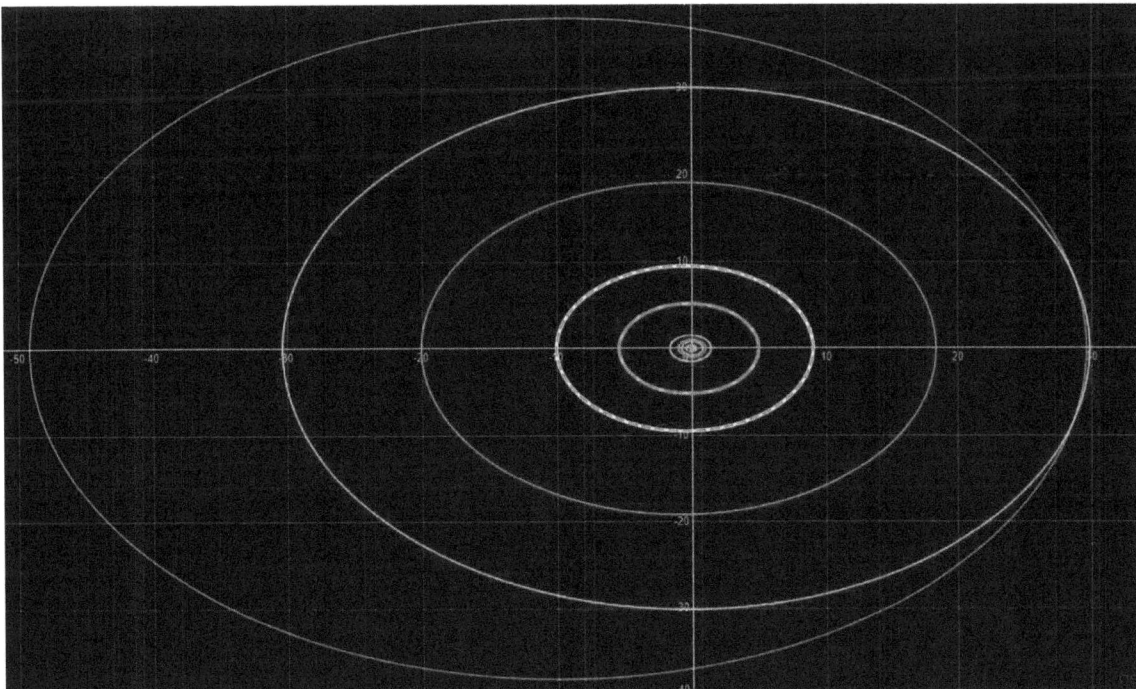

3. Functions in Cylindrical Coordinates in 3D:

$r = \sin(\theta + z^2)\ \{0 \leq \theta \leq 2\pi\}$

$r = \sin(\theta^2 + z)\ \{0 \leq \theta \leq 2\pi\}$

$r = \sin(\theta + z)\ \{0 \leq \theta \leq 2\pi\}$

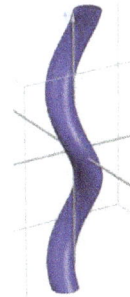

$r = \sin^2(\theta\sqrt{z})\ \{0 \leq \theta \leq 2\pi\}$

$r = \sin^2(z\theta)\ \{0 \leq \theta \leq 2\pi\}$

$r = \sin^2(\theta)$ and $r = z\sin^2(\theta)$ $\{0 \leq \theta \leq 2\pi\}$

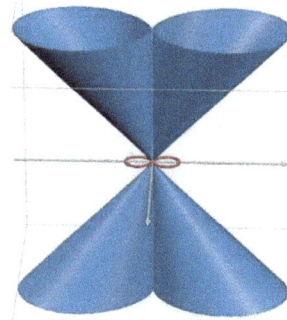

2. Roses in 3D

$r = 2z\cos(3\theta)$
$\{0 \leq \theta \leq 2\pi\}$
$\{-1 \leq z \leq 1\}$

$r = 2\cos(3\theta)$
$\{0 \leq \theta \leq 2\pi\}$
$\{-1 \leq z \leq 1\}$

$r = 2\sin(z)\cos(3\theta)$
$\{0 \leq \theta \leq 2\pi\}\quad \{-\pi \leq z \leq \pi\}$

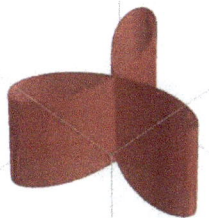

91

$$r = 2\sin(z^2)\cos(3\theta)$$
$$\{0 \leq \theta \leq 2\pi\}$$
$$\{-\pi \leq z \leq \pi\}$$

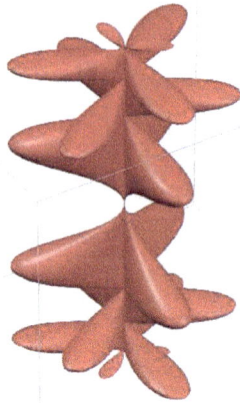

$$r = 2\sin(z^3)\cos(3\theta)$$
$$\{0 \leq \theta \leq 2\pi\}$$
$$\{-\pi \leq z \leq \pi\}$$

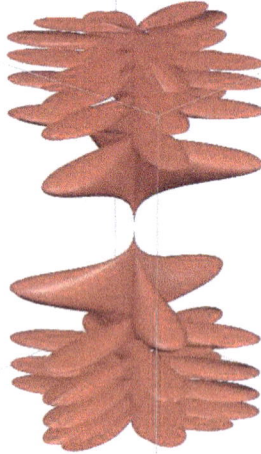

$$r = 2\sin(z^3+1)\cos(3\theta)$$
$$\{0 \leq \theta \leq 2\pi\}$$
$$\{-\pi \leq z \leq \pi\}$$

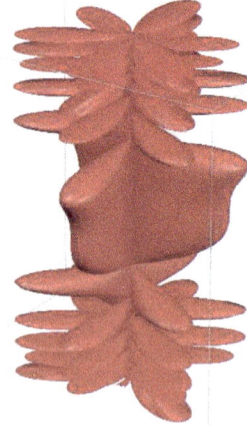

$$r = 2\sin(\sqrt{z})\cos(3\theta)$$
$$\{0 \leq \theta \leq 2\pi\}$$
$$\{0 \leq z \leq 4\pi^2\}$$

$$r = 2\sin(\sqrt{z}+1)\cos(3\theta)$$
$$\{0 \leq \theta \leq 2\pi\}$$
$$\{0 \leq z \leq 4\pi^2\}$$

$$r = 2\sin(\sqrt{z+1})\cos(3\theta)$$
$$\{0 \leq \theta \leq 2\pi\}$$
$$\{-1 \leq z \leq 4\pi^2 - 1\}$$

$$r = 2\sin(4\theta) \ \text{and} \ r = 2\cos(4\theta)$$
$$\{0 \leq \theta \leq 2\pi\}$$
$$\{-1 \leq z \leq 1\}$$

$$r = 2z\cos(4\theta)$$
$$\{0 \leq \theta \leq 2\pi\}$$
$$\{-1 \leq z \leq 1\}$$

$$r = 2\sin(z)\cos(4\theta)$$
$$\{0 \leq \theta \leq 2\pi\}$$
$$\{-\pi \leq z \leq \pi\}$$

$r = 2\sin(z^2 + z + 1)\cos(3\theta)$
$\{0 \le \theta \le 2\pi\}$
$\{-\pi \le z \le \pi\}$

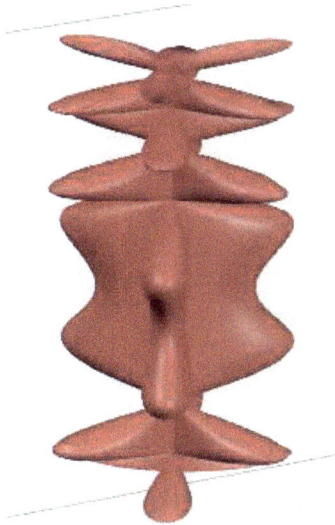

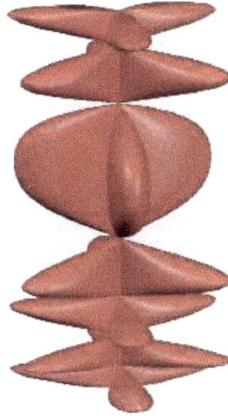

$r = 2\sin(z^2 - z - 1)\cos(3\theta)$
$\{0 \le \theta \le 2\pi\}$
$\{-\pi \le z \le \pi\}$

$r = 2\sin(z^2 - z + 1)\cos(3\theta)$
$\{0 \le \theta \le 2\pi\}$
$\{-\pi \le z \le \pi\}$

$r = 2\sin(z^2)\cos(4\theta)$
$\{0 \le \theta \le 2\pi\}$
$\{-\pi \le z \le \pi\}$

$r = 2\sin(z^2 + z + 1)\cos(4\theta)$
$\{0 \le \theta \le 2\pi\}$
$\{-\pi \le z \le \pi\}$

$r = 2\sin(z^2 - z - 1)\cos(4\theta)$
$\{0 \le \theta \le 2\pi\}$
$\{-\pi \le z \le \pi\}$

$$r = 2\sin(z^2 - z)\cos(4\theta)$$
$$\{0 \leq \theta \leq 2\pi\}$$
$$\{-\pi \leq z \leq \pi\}$$

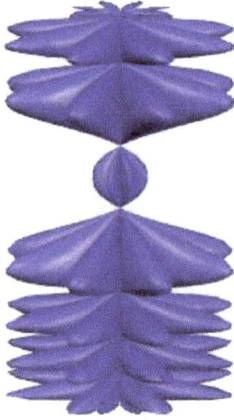

$$r = 2\sin(z^3 - z)\cos(4\theta)$$
$$\{0 \leq \theta \leq 2\pi\}$$
$$\{-\frac{\pi}{2} \leq z \leq \frac{\pi}{2}\}$$

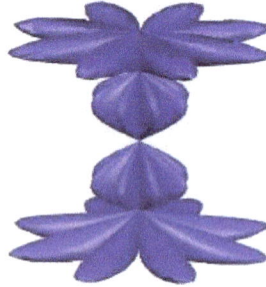

$$r = 2\sin(z^3 - z - 1)\cos(4\theta)$$
$$\{0 \leq \theta \leq 2\pi\}$$
$$\{-\frac{2\pi}{3} \leq z \leq \frac{2\pi}{3}\}$$

$$r = 2\sin(z^3 - 2z^2 - z + 2)\cos(4\theta)$$
$$\{0 \leq \theta \leq 2\pi\}$$
$$\{-\frac{\pi}{2} \leq z \leq \frac{5\pi}{6}\}$$

$$r = 2\sin(z^4 - z)\cos(4\theta)$$
$$\{0 \leq \theta \leq 2\pi\}$$
$$\{-\frac{\pi}{2} \leq z \leq \frac{5\pi}{6}\}$$

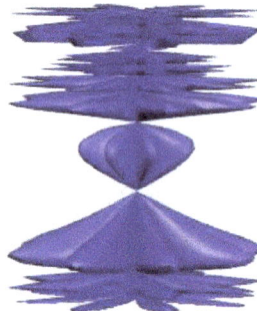

$$r = 2\sin(z^4 - 1)\cos(4\theta)$$
$$\{0 \leq \theta \leq 2\pi\}$$
$$\{-\frac{\pi}{2} \leq z \leq \frac{5\pi}{6}\}$$

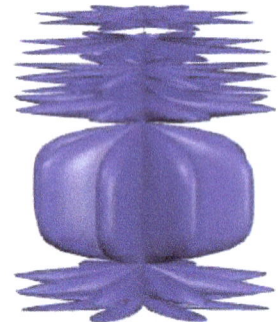

$$r = 2\sin(z^4 - 1)\cos(4\theta)$$
$$\{0 \leq \theta \leq 2\pi\}$$
$$\{-\frac{\pi}{2} \leq z \leq \frac{5\pi}{6}\}$$

$$r = 2\sin(z^4 - z^3)\cos(4\theta)$$
$$\{0 \le \theta \le 2\pi\}$$
$$\{-\frac{\pi}{2} \le z \le \frac{5\pi}{6}\}$$

$$r = 2\sin(\sqrt{z})\cos(4\theta)$$
$$\{0 \le \theta \le 2\pi\}$$
$$\{0 \le z \le 4\pi^2\}$$

$$r = 2\sin(\sqrt{z} - 1)\cos(4\theta)$$
$$\{0 \le \theta \le 2\pi\}$$
$$\{0 \le z \le 4\pi^2\}$$

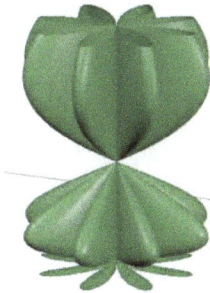

$$r = 2\sin(\sqrt{z} + 1)\cos(4\theta)$$
$$\{0 \le \theta \le 2\pi\}$$
$$\{0 \le z \le 4\pi^2\}$$

$$r = 2\sin(e^z)\cos(4\theta)$$
$$\{0 \le \theta \le 2\pi\}$$
$$\{-\frac{\pi}{e} \le z \le \frac{\pi}{e}\}$$

$$r = 2\sin(e^{z^2})\cos(4\theta)$$
$$\{0 \le \theta \le 2\pi\}$$
$$\{-\frac{\pi}{e} \le z \le \frac{\pi}{e}\}$$

$$r = 2\sin(\ln(z))\cos(4\theta)$$
$$\{0 \le \theta \le 2\pi\}$$
$$\{0 \le z \le e^{2\pi}\}$$

95

1. Functions and Their Graphs in Spherical Coordinates

$\rho = 2\sin(\theta\cos(\phi))$ $\{-2\pi \leq \theta \leq 2\pi\}$ $\{0 \leq \phi \leq \pi\}$	$\rho = 2 + \sin(\theta\cos(\phi))$ $\{-2\pi \leq \theta \leq 2\pi\}$ $\{0 \leq \phi \leq 2\pi\}$	$\rho = 2(1 + \sin(\theta + \cos(\phi))$ $\{-\pi \leq \theta \leq \pi\}$ $\{0 \leq \phi \leq 2\pi\}$

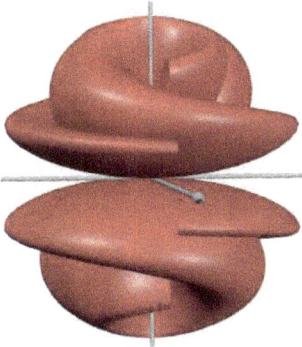

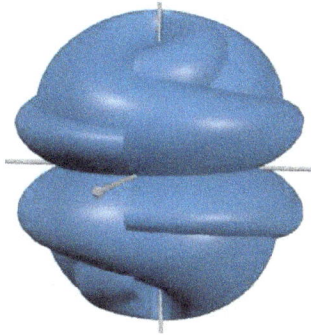

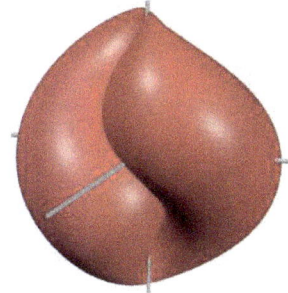

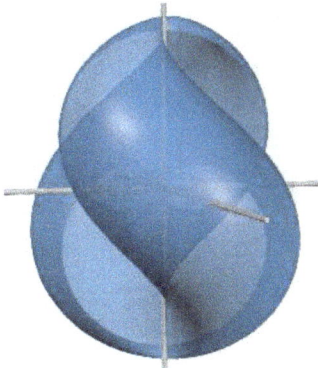

$\rho = 2(1 + \sin(\theta + \cos(\phi))$ $\{0 \leq \theta \leq \pi\}$ $\{0 \leq \phi \leq 2\pi\}$	$\rho = 2\sin(2\theta + \cos(\phi))$ $\{0 \leq \theta \leq \pi\}$ $\{0 \leq \phi \leq 2\pi\}$	$\rho = 2\sin(3\theta + \cos(\phi))$ $\{0 \leq \theta \leq \pi\}$ $\{0 \leq \phi \leq 2\pi\}$

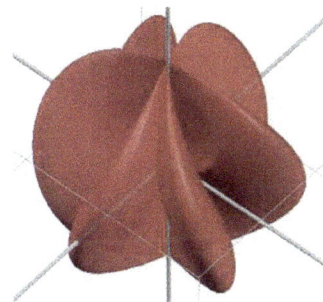

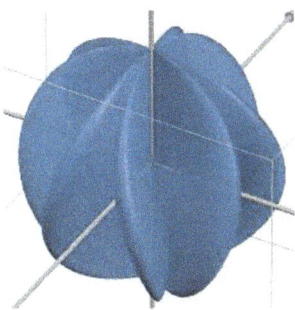

$\rho = 2\sin(4\theta + \cos(\phi))$ $\{0 \leq \theta \leq \pi\}$ $\{0 \leq \phi \leq 2\pi\}$	$\rho = 2(1 + \sin(2\theta + \csc(\phi))$ $\{0 \leq \theta \leq 2\pi\}$ $\{\frac{\pi}{9} \leq \phi \leq \frac{7\pi}{9}\}$	$\rho = 2(1 + \sin(2\theta + \csc(\phi))$ $\{0 \leq \theta \leq 2\pi\}$ $\{\frac{\pi}{18} \leq \phi \leq \frac{17\pi}{18}\}$

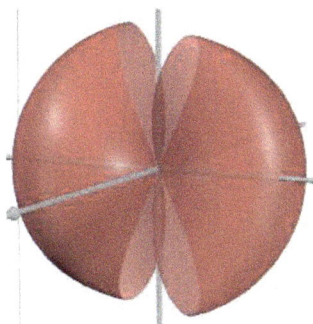

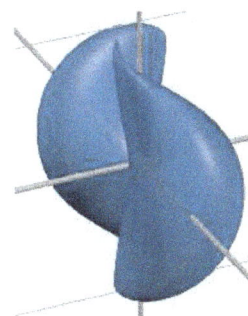

$\rho = 2(2 + \sin(2\theta + \cos(\phi)))$
$\{0 \le \theta \le \pi\}$ $\{0 \le \phi \le 2\pi\}$

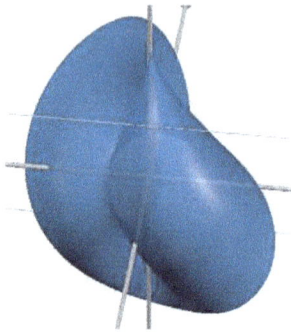

$\rho = 2(3 + \sin(\theta\cos(\phi)))$
$\{-2\pi \le \theta \le 2\pi\}$
$\{0 \le \phi \le 2\pi\}$

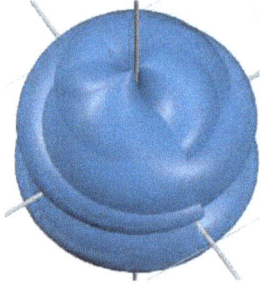

$\rho = \sin(\theta)\cos(\phi)$
$\{0 \le \theta \le 2\pi\}$
$\{0 \le \phi \le \pi\}$

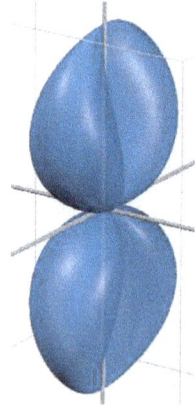

$\rho = \sin(\theta) + \cos(\phi)$
$\{0 \le \theta \le 2\pi\}$
$\{0 \le \phi \le 2\pi\}$

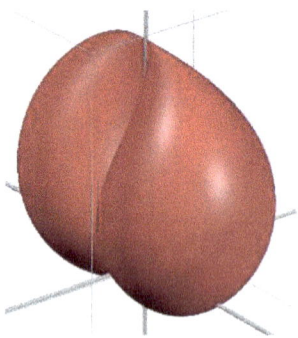

$\rho = \sin(\theta) + \cos(\phi) + 2$
$\{0 \le \theta \le 2\pi\}$
$\{0 \le \phi \le \pi\}$

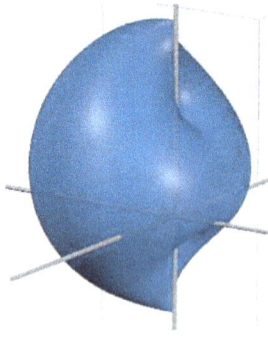

$\rho = \sin^2(\theta) + \cos(\phi)) + 4$
$\{0 \le \theta \le 2\pi\}$
$\{0 \le \phi \le 2\pi\}$

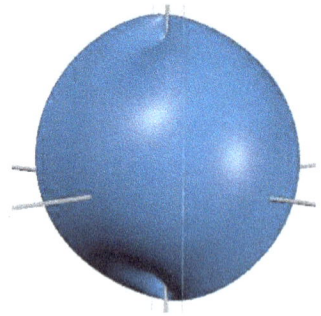

$\rho = \sin^2(\theta) + \cos^2(\phi) + 4$
$\{0 \le \theta \le 2\pi\}$ $\{0 \le \phi \le \pi\}$

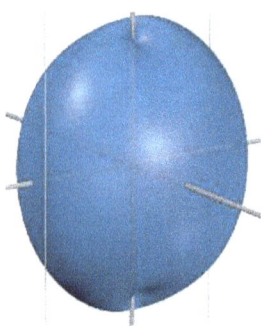

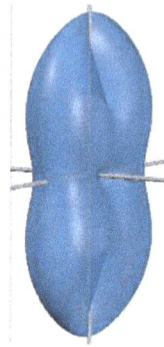

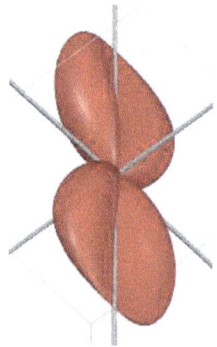

$\rho = \sin^2(\theta) + \cos^2(\phi)$
$\{0 \le \theta \le 2\pi\}$ $\{0 \le \phi \le \pi\}$

$\rho = \sin^2(\theta)\cos^2(\phi)$
$\{0 \le \theta \le 2\pi\}$ $\{0 \le \phi \le \pi\}$

98

$$\rho = \sin^2(\theta)\cos^2(\phi) + 1$$
$$\{0 \leq \theta \leq 2\pi\}$$
$$\{0 \leq \phi \leq \pi\}$$

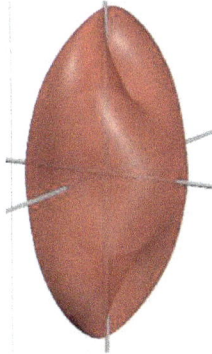

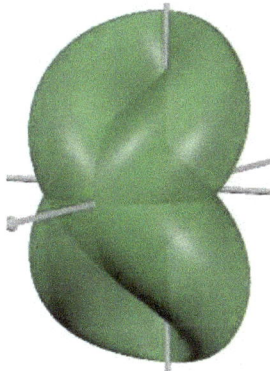

$$\rho = \sin(\theta^2)\cos^2(\phi)$$
$$\{0 \leq \theta \leq 2\pi\}$$
$$\{0 \leq \phi \leq 2\pi\}$$

$$\rho = \theta$$
$$\{0 \leq \theta \leq 2\pi\}$$
$$\{0 \leq \phi \leq \pi\}$$

$$\rho = \sin(\theta + \phi)$$
$$\{0 \leq \theta \leq 2\pi\}$$
$$\{0 \leq \phi \leq \pi\}$$

$$\rho = 2$$
$$\{0 \leq \theta \leq 2\pi\}$$
$$\{0 \leq \phi \leq 2\pi\}$$

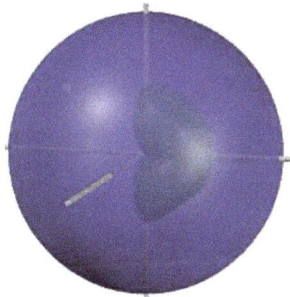

$$\rho = 1 + \frac{1}{2}\sin(3\theta)$$
$$\{0 \leq \theta \leq 2\pi\}$$
$$\{0 \leq \phi \leq \pi\}$$

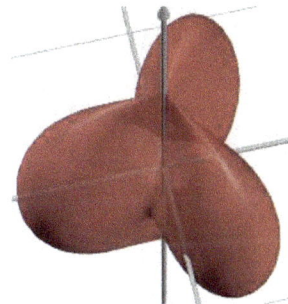

$$\rho = 1 + \frac{1}{2}\sin(4\theta)$$
$$\{0 \leq \theta \leq 2\pi\}$$
$$\{0 \leq \phi \leq \pi\}$$

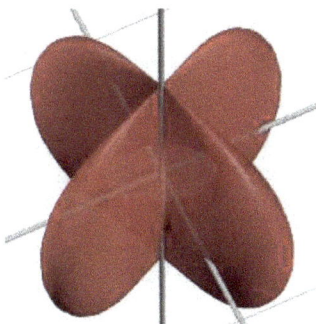

$$\rho = 1 + \frac{1}{2}\sin\left(\frac{1}{3}\theta\right)$$
$$\{0 \leq \theta \leq 2\pi\}$$
$$\{0 \leq \phi \leq \pi\}$$

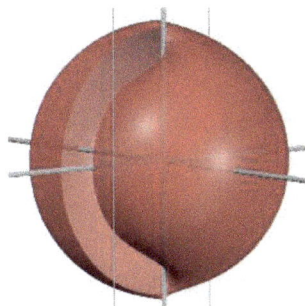

$$\rho = 1 + \frac{1}{2}\sin(6\theta)\sin(4\phi)$$
$$\{0 \leq \theta \leq 2\pi\}$$
$$\{0 \leq \phi \leq \pi\}$$

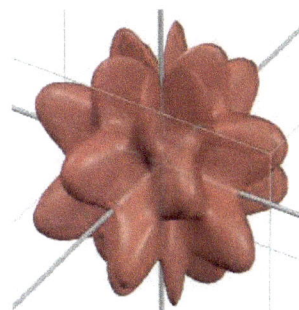

$$\rho = 1 + \frac{1}{7}\sin(6\theta)\sin(4\phi)$$
$$\{0 \leq \theta \leq 2\pi\}$$
$$\{0 \leq \phi \leq \pi\}$$

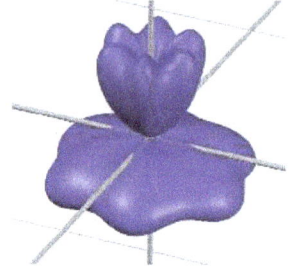

$$\rho = 1 + \frac{1}{7}\sin(6\theta) + \sqrt{\theta}$$
$$\{0 \leq \theta \leq 2\pi\}$$
$$\{0 \leq \phi \leq \pi\}$$

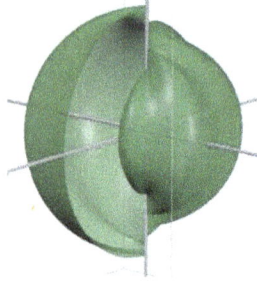

$$\rho = 1 + \frac{1}{7}\sin(6\theta) + \sin(4\phi)$$
$$\{0 \leq \theta \leq 2\pi\}$$
$$\{0 \leq \phi \leq \pi\}$$

$$\rho = 4\sin(\phi) - 2$$
$$\{0 \leq \theta \leq 2\pi\} \quad \{0 \leq \phi \leq \pi\}$$

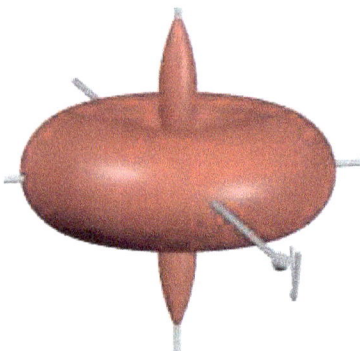

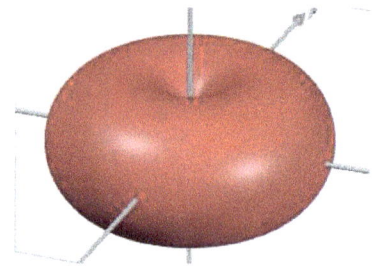

$$\rho = 4\sin(\phi) - 3$$
$$\{0 \leq \theta \leq 2\pi\} \quad \{0 \leq \phi \leq \pi\}$$

$$\rho = 4\sin(\phi)$$
$$\{0 \leq \theta \leq 2\pi\} \quad \{0 \leq \phi \leq \pi\}$$

$$\rho = 4\sin(\theta\sin(\phi))$$
$$\{0 \leq \theta \leq 2\pi\}$$
$$\{0 \leq \phi \leq 2\pi\}$$

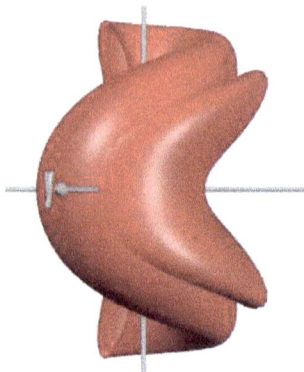

$$\rho = 4(1 + \sin(\theta)\sin(\phi))$$
$$\rho = 4\sin(\theta)\cos(\phi)$$
$$\rho = 4\sin\left(\theta + [0, \mp\frac{\pi}{2}, \pi\right)\sin(\phi)$$
$$\{0 \leq \theta \leq 2\pi\} \quad \{0 \leq \phi \leq \pi\}$$

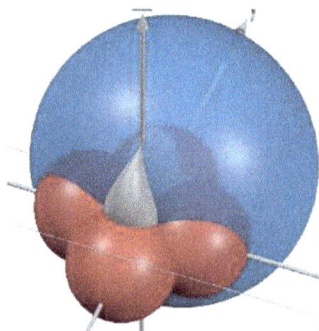

$$\rho = 4\sin(\theta)\sin(\phi) \text{ and } \rho = 1$$
$$\{0 \leq \theta \leq 2\pi\}$$
$$\{0 \leq \phi \leq \pi\}$$

$\rho = 4\theta\phi$
$\{0 \leq \theta \leq 2\pi\}$
$\{0 \leq \phi \leq 2\pi\}$

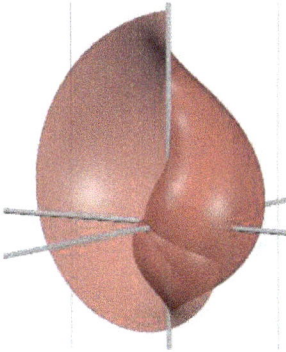

$\rho = 4\cos(\theta\phi)$
$\{0 \leq \theta \leq 2\pi\}$
$\{0 \leq \phi \leq 2\pi\}$

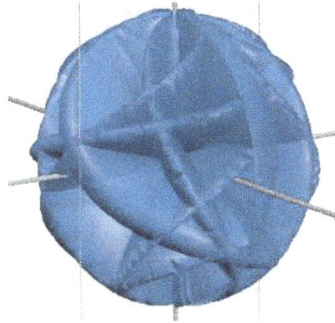

$\rho = 4\cos(2\theta\phi)$
$\{0 \leq \theta \leq 2\pi\}$
$\{0 \leq \phi \leq 2\pi\}$

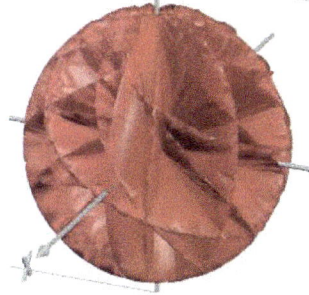

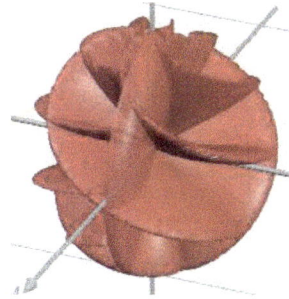

$\rho = 4\sin(3\theta^2\phi)$
$\{0 \leq \theta \leq 2\pi\}$
$\{0 \leq \phi \leq 2\pi\}$

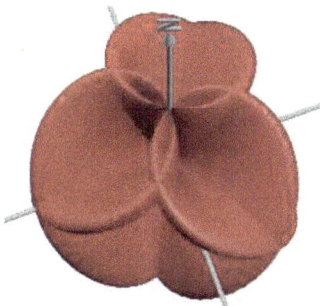

$\rho = 4\cos(3\theta\phi)$
$\{0 \leq \theta \leq 2\pi\}$
$\{0 \leq \phi \leq 2\pi\}$

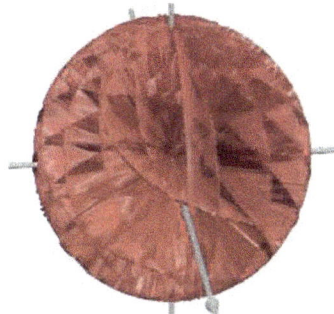

$\rho = 4\cos(2\theta\phi)$
$\{0 \leq \theta \leq 2\pi\}$
$\{0 \leq \phi \leq \pi\}$

$\rho = 4\sin(3(\theta + \phi))$
$\{0 \leq \theta \leq 2\pi\}$
$\{0 \leq \phi \leq 2\pi\}$

$\rho = 4\sin(4(\theta + \phi))$
$\{0 \leq \theta \leq 2\pi\}$
$\{0 \leq \phi \leq 2\pi\}$

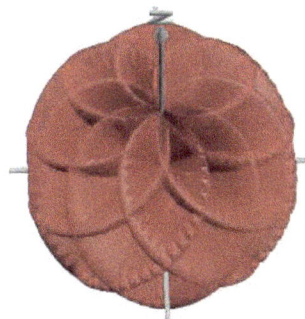

$\rho = 4\sin(4\sqrt{\theta + \phi})$
$\{0 \leq \theta \leq 2\pi^2\}$
$\{0 \leq \phi \leq 2\pi^2\}$

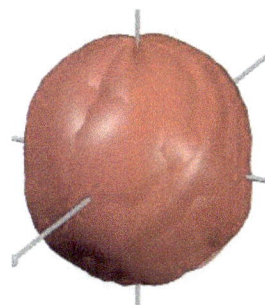

$$\rho = 4\cos(\theta)\phi - 3$$
$$\{0 \leq \theta \leq 2\pi\}$$
$$\{0 \leq \phi \leq 2\pi\}$$

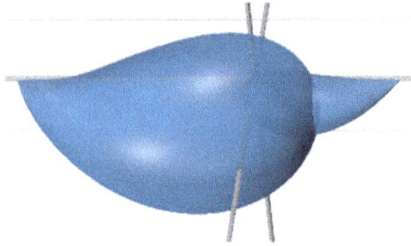

$$\rho = 4\sin(\theta)(\phi)$$
$$\{0 \leq \theta \leq 2\pi\}$$
$$\{0 \leq \phi \leq 2\pi\}$$

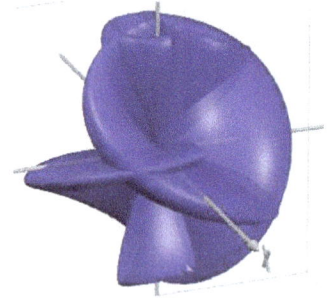

$$\rho = 4\sin(\theta\phi)$$
$$\{0 \leq \theta \leq 2\pi\}$$
$$\{0 \leq \phi \leq \pi\}$$

$$\rho = 4\cos(4(\theta + \phi))$$
$$\{0 \leq \theta \leq 2\pi\}$$
$$\{0 \leq \phi \leq 2\pi\}$$

$$\rho = 4\cos(4\sqrt{\theta + \phi})$$
$$\{0 \leq \theta \leq 2\pi\}$$
$$\{0 \leq \phi \leq 2\pi\}$$

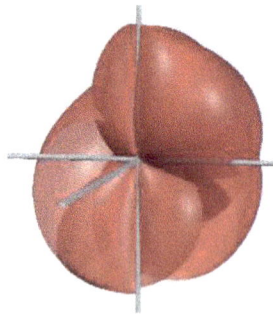

$$\rho = -4\sin(4\sqrt{\theta + \phi})$$
$$\{0 \leq \theta \leq 2\pi\}$$
$$\{0 \leq \phi \leq 2\pi\}$$

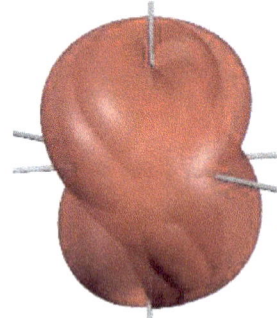

$$\rho = 2\sin(\phi) + 5$$
$$r = 7\sin(z^2 - z)\cos(2\theta)\{5 \leq z \leq 6\}$$
$$\{0 \leq \theta \leq 2\pi\} \quad \{0 \leq \phi \leq \pi\}$$

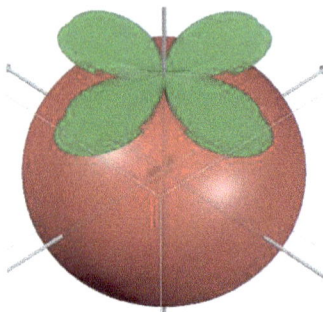

$$\rho = 2\sin\left(\frac{\phi}{2}\right)$$
$$\{0 \leq \theta \leq 2\pi\}$$
$$\{0 \leq \phi \leq \pi\}$$

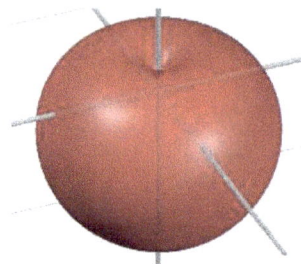

$$\rho = 5\cos(\theta) + 2\sin\left(\frac{\phi}{2}\right)$$
$$\{0 \leq \theta \leq 2\pi\}$$
$$\{0 \leq \phi \leq \pi\}$$

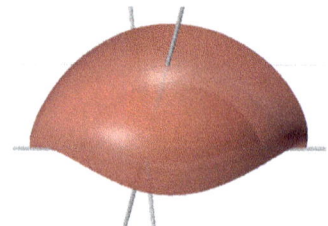

2. Gaussian Functions in Spherical Coordinates

12 -17

$$\rho = \frac{1}{\sqrt{2\pi}} e^{-\sin(\theta)\sin(\phi)}$$

$$\rho = \frac{1}{\sqrt{2\pi}} e^{\sin(\theta)\sin(\phi)}$$

$$\rho = \frac{1}{\sqrt{2\pi}} e^{\sin\left(\theta+\frac{\pi}{2}\right)\sin(\phi)}$$

$$\rho = \frac{1}{\sqrt{2\pi}} e^{\sin\left(\theta-\frac{\pi}{2}\right)\sin(\phi)}$$

$$\rho = \frac{1}{\sqrt{2\pi}} e^{\sin\left(\theta-\frac{\pi}{2}\right)\sin\left(\phi+\frac{\pi}{2}\right)}$$

$$\rho = \frac{1}{\sqrt{2\pi}} \ln\left(1 + \sin\left(\theta - \frac{\pi}{2}\right)\sin\left(\phi + \frac{\pi}{2}\right)\right)$$

$$0 \le \theta \le 2\pi, 0 \le \phi \le \pi$$

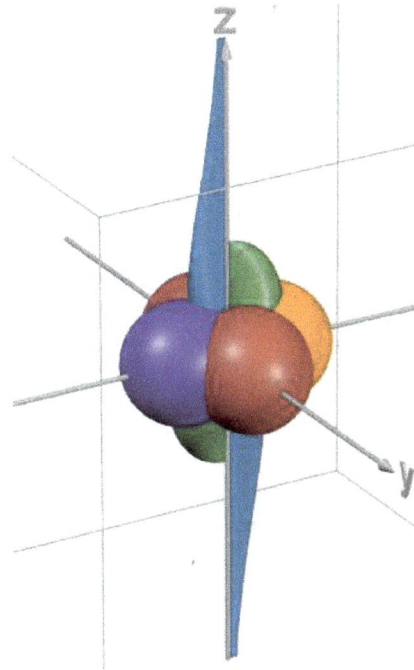

7, 9

$$\rho = \frac{1}{\sqrt{2\pi}} e^{-\sin(\theta+\phi)^2} - 1$$

$$\rho = 1 - \frac{1}{\sqrt{2\pi}} e^{-(\sin(\theta+\phi))^2}$$

$$0 \le \theta \le 2\pi, 0 \le \phi \le \pi$$

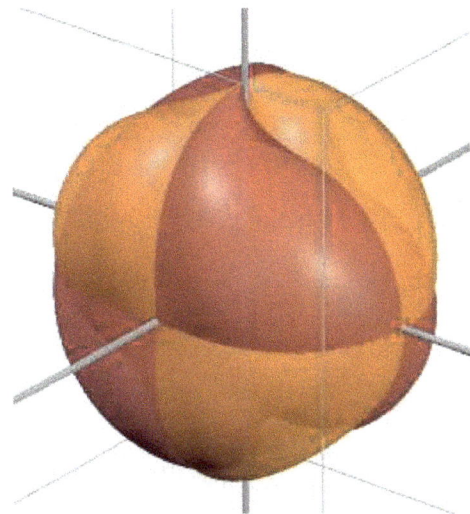

$$\boxed{2,4,19}$$

$$\rho = \frac{1}{\sqrt{2\pi}} e^{-\theta}$$

$$\rho = \frac{2}{\sqrt{2\pi}} e^{-\frac{(\theta+\phi)^2}{2}}$$

$$\rho = \frac{1}{\sqrt{2\pi}} \ln(1 + \sin(\theta)\sin(\phi))$$

$$0 \le \theta \le 2\pi, 0 \le \phi \le \pi$$

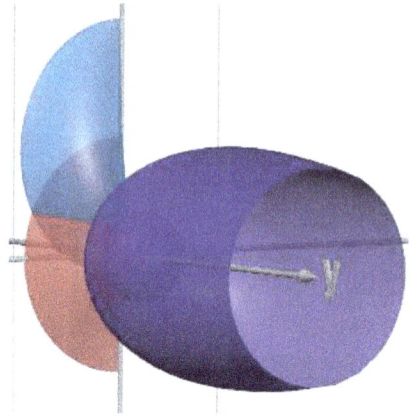

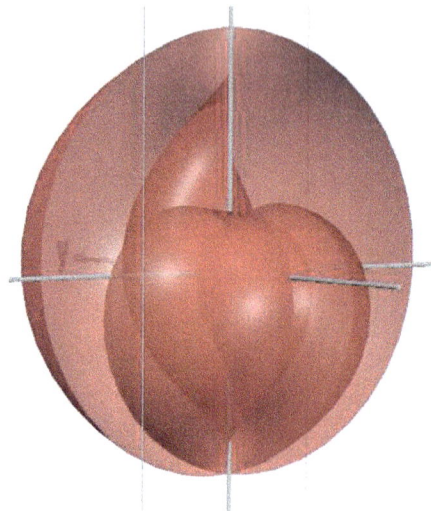

$$\boxed{23}$$

$$\rho = \frac{1}{\sqrt{2\pi}} e^{-\tan^{-1}(\theta)\tan^{-1}(\phi)}$$

$$0 \le \theta \le 2\pi, 0 \le \phi \le \pi$$

$$\rho = \frac{1}{\sqrt{2\pi}} e^{-\sin^{-1}\left(\theta-\frac{1}{2}\right)\sin^{-1}\left(\phi+\frac{1}{2}\right)} \quad \boxed{24 \text{ G}}$$

$$\rho = \frac{1}{\sqrt{2\pi}} e^{\sin^{-1}\left(\theta-\frac{1}{2}\right)\sin^{-1}\left(\phi+\frac{1}{2}\right)}$$

$$\rho = \frac{-1}{\sqrt{2\pi}} e^{\sin^{-1}\left(\theta-\frac{1}{2}\right)\sin^{-1}\left(\phi+\frac{1}{2}\right)}$$

$$\{-0.5 \le \theta \le 1.5, -1.5 \le \phi \le 0.5\}$$

$$\rho = \frac{1}{\sqrt{2\pi}} e^{-\sin^{-1}\left(\theta+\frac{1}{2}\right)\sin^{-1}\left(\phi-\frac{1}{2}\right)}$$

$$\rho = \frac{1}{\sqrt{2\pi}} e^{\sin^{-1}\left(\theta+\frac{1}{2}\right)\sin^{-1}\left(\phi-\frac{1}{2}\right)}$$

$$\rho = \frac{-1}{\sqrt{2\pi}} e^{\sin^{-1}\left(\theta+\frac{1}{2}\right)\sin^{-1}\left(\phi-\frac{1}{2}\right)}$$

$$-1.5 \le \theta \le 0.5, -0.5 \le \phi \le 1.5$$

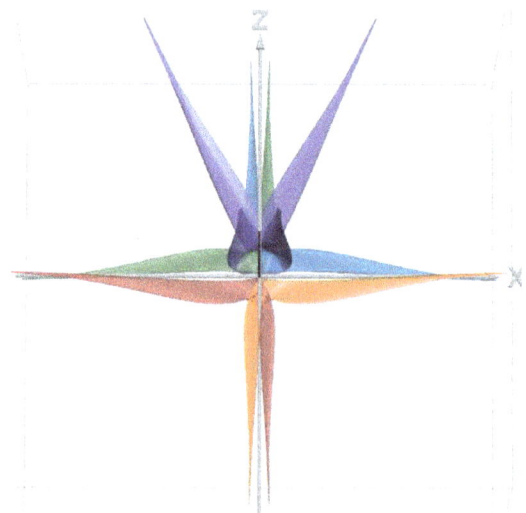

104

Continue on the Spherical Coordinates

3.   The number Six in Spherical Coordinates

The number six holds significant symbolism across various cultures and belief systems, often associated with harmony, balance, and perfection. In numerology, six is considered several nurturing, responsibility, and compassion. People see six as a number representing family, love, and domestic affairs. Additionally, in some cultures, six is believed to symbolize harmony between the material and spiritual worlds.

One common myth related to the number six is its association with the concept of the "mark of the beast" from the Book of Revelation in the Bible. The number 666 is often feared and regarded as a sign of evil or the devil. This myth has led to a fear of the number six in some individuals, causing anxiety and superstition.

Understanding the origins and interpretations of such beliefs is essential to overcome the fear associated with the myth of the number six. By educating oneself about numerology, symbolism, and cultural contexts, one can gain a deeper appreciation for the positive aspects of the number six. Embracing the idea of balance, harmony, and responsibility that six represents can help shift the perception from fear to acceptance.

Practicing mindfulness and rational thinking can benefit those overwhelmed by the fear of the myth surrounding the number six. By focusing on the positive traits associated with the number and challenging negative beliefs, individuals can gradually overcome their anxiety and superstitions. Engaging in activities that promote balance and harmony in one's life can also help change the perception of the number six from a source of fear to a symbol of positivity and strength.

In conclusion, six is a potent symbol with various meanings and interpretations. By understanding the significance of six, acknowledging the myths surrounding it, and working towards overcoming irrational fears, individuals can embrace the positive qualities of this number and incorporate them into their lives for a sense of balance and harmony.

The following two pages show how six can be used as a scaler to change a sphere into various shapes.

$\rho = 6$
$\{-5 \le x, y, z \le 5\}$
$0 \le \theta \le 2\pi, 0 \le \phi \le \pi$

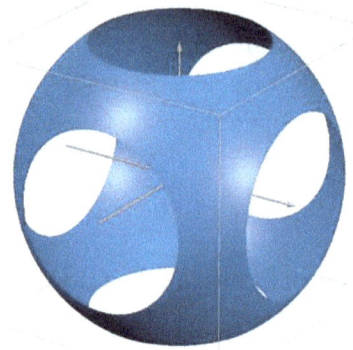

$\rho = 6 \sin(\theta) \sin(\phi)$
$\{-5 \le x, y, z \le 5\}$
$0 \le \theta \le 2\pi, 0 \le \phi \le \pi$

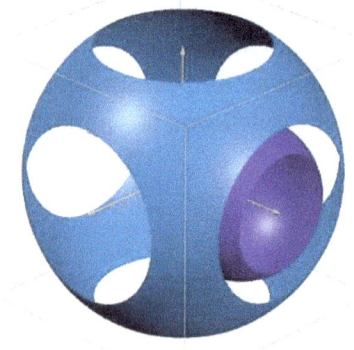

$\rho = 6 \sin(6\theta) \sin(\phi)$
$\{-5 \le x, y, z \le 5\}$
$0 \le \theta \le 2\pi, 0 \le \phi \le \pi$

$\rho = 6 \sin(6\theta) \sin(6\phi)$
$\{-5 \le x, y, z \le 5\}$
$0 \le \theta \le 2\pi, 0 \le \phi \le \pi$

$\rho = 6(1 + \sin(6\theta) \sin(6\phi))$
$\{-9 \le x, y, z \le 9\}$
$0 \le \theta \le 2\pi, 0 \le \phi \le \pi$

$\rho = 6(1 + \sin(6\theta) \sin(6\phi))$
$\{-14 \le x, y, z \le 14\}$
$0 \le \theta \le 2\pi, 0 \le \phi \le \pi$

$\rho = 6 \sin(6\theta) \sin(6\phi)$
$\{-12 \le x, y, z \le 12\}$
$0 \le \theta \le 2\pi, 0 \le \phi \le \pi$

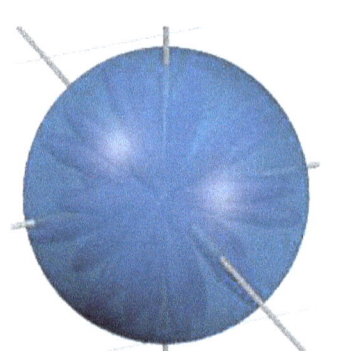

$\rho = 6 \sin(66\theta) \sin(6\phi)$
$\{-12 \le x, y, z \le 12\}$
$0 \le \theta \le 2\pi, 0 \le \phi \le \pi$

$\rho = 6 \sin(66\theta) \sin(66\phi)$
$\{-9 \le x, y, z \le 9\}$
$0 \le \theta \le 2\pi, 0 \le \phi \le \pi$

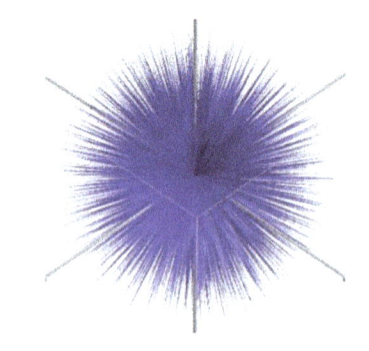

$$\rho = 6\sin(666\theta)\sin(66\phi)$$
$$\{-7 \le x, y, z \le 7\}$$
$$0 \le \theta \le 2\pi, 0 \le \phi \le \pi$$

$$\rho = 6\sin(666\theta)\sin(666\phi)$$
$$\{-7 \le x, y, z \le 7\}$$
$$0 \le \theta \le 2\pi, 0 \le \phi \le \pi$$

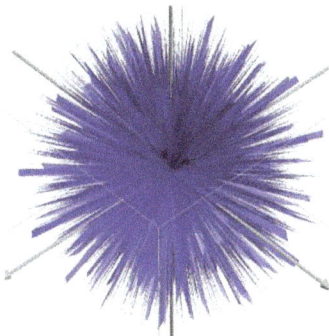

$$\rho = 6\sin\left(\frac{1}{6}\theta\right)\sin(666\phi)$$
$$\{-9 \le x, y, z \le 9\}$$
$$0 \le \theta \le 2\pi, 0 \le \phi \le \pi$$

$$\rho = 6\sin\left(\frac{1}{6}\theta\right)\sin(6\phi)$$
$$\{-12 \le x, y, z \le 12\}$$
$$0 \le \theta \le 2\pi, 0 \le \phi \le \pi$$

$$\rho = 6\sin\left(\frac{1}{6}\theta\right)\sin\left(\frac{1}{6}\phi\right)$$
$$\{-4 \le x, y, z \le 4\}$$
$$0 \le \theta \le 2\pi, 0 \le \phi \le \pi$$

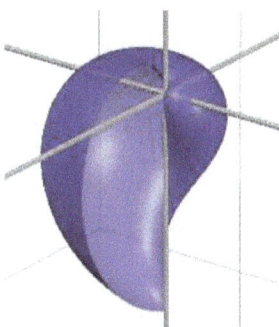

$$\rho = 6\sin(6\theta)\sin\left(\frac{1}{6}\phi\right)$$
$$\{-7 \le x, y, z \le 7\}$$
$$0 \le \theta \le 2\pi, 0 \le \phi \le \pi$$

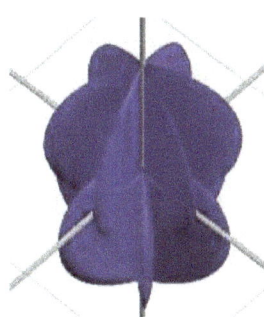

$$\rho = 6\sin(6\theta)\sin\left(\frac{1}{6}\phi\right) + \frac{1}{666 + \theta^2}$$
$$\{-12 \le x, y, z \le 12\}$$
$$0 \le \theta \le 2\pi, 0 \le \phi \le \pi$$

$$\rho = 6\sin(6\theta)\sin\left(\frac{1}{6}\phi\right) + \frac{6}{\sec(\phi)}$$
$$\{-12 \le x, y, z \le 12\}$$
$$0 \le \theta \le 2\pi, 0 \le \phi \le \pi$$

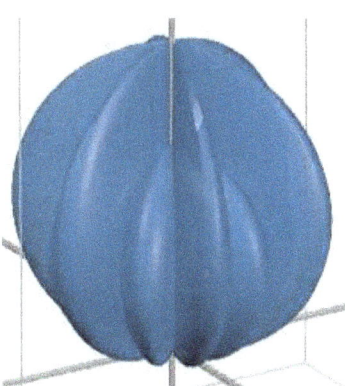

$$\rho = 6\sin(6\theta)\sin\left(\frac{1}{6}\phi\right) + \frac{6}{\tan(\phi)}$$
$$\{-19 \le x, y, z \le 19\}$$
$$0 \le \theta \le 2\pi, 0 \le \phi \le \pi$$

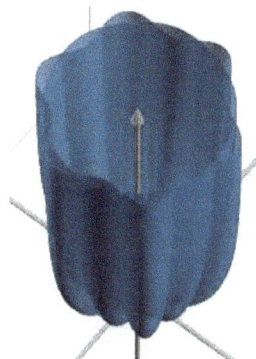

107

VI.   Vectors in Cartesian Coordinates

1. Unit Vectors: i, j, k
2. Vectors and Dot Product
3. Operations (Form New Vectors)
    a. Addition
    b. Subtraction
    c. Opposite
    d. Cross Product
4. Lines and Planes

1.   Unit Vectors: i, j, k

| Unit Vector i | Unit Vector j | Unit Vector k |

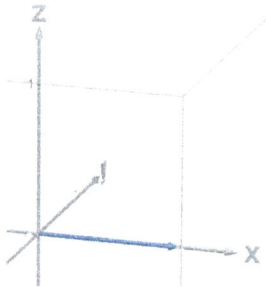 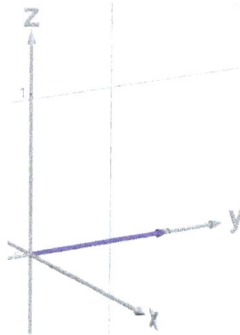 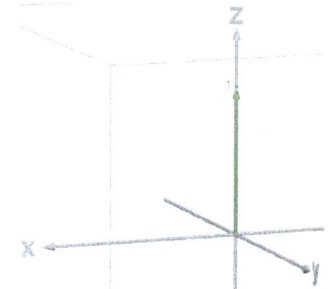

2.   Vectors and Dot Product

| Vector $u = ((1,1,1), (2,3,5))$ | Vector $v = ((0,3,2), (5,3,-1))$ | Dot Product u and v $$u \cdot v = -7$$ |

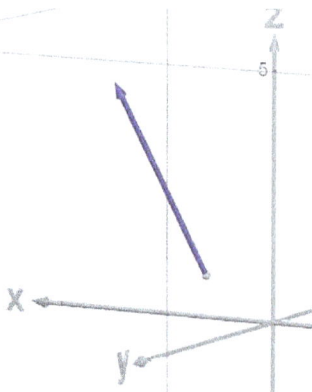 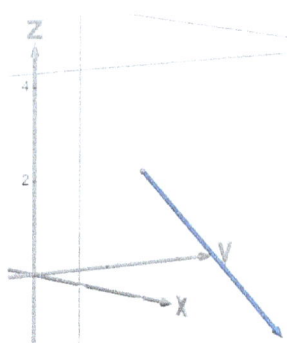 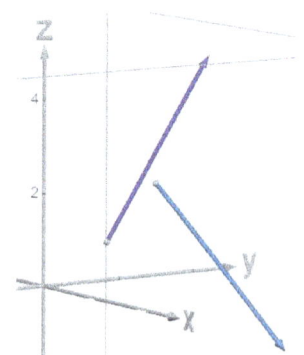

3. Operations (Form New Vectors)

Sum of u and v

$$u + v = ((1,1,1), (7,3,2))$$

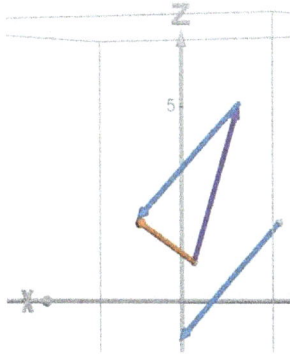

Difference of u and v

$$u - v = ((1,1,1), (-3,3,8))$$

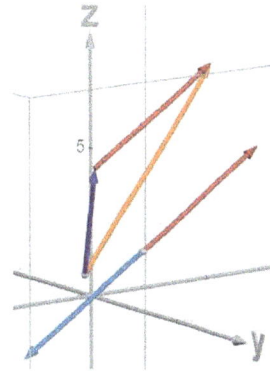

Opposite of v

$$-v = ((0,3,2), (-5,3,5))$$

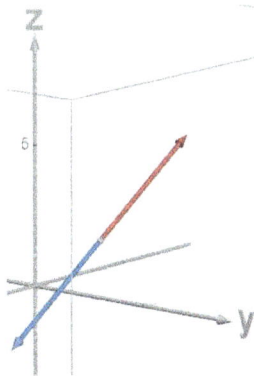

Cross Product of u and v

$$u \times v = ((1,1,1), (-5,24,-9))$$

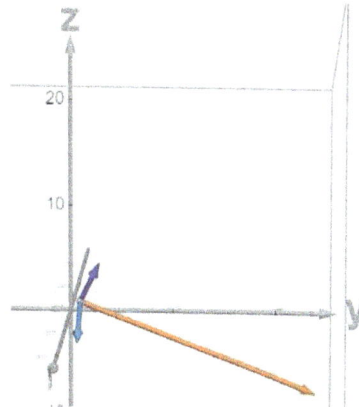

The length of u: $|u| = \sqrt{(2-1)^2 + (3-1)^2 + (5-1)^2} = \sqrt{21}$

The length of v: $|v| = \sqrt{(5-0)^2 + (3-3)^2 + (-1-2)^2} = \sqrt{34}$

The length of $u \times v$: $|u \times v| = \sqrt{(-5-1)^2 + (24-1)^2 + (-9-1)^2} = \sqrt{665}$

The angle between u and v: $\theta = \sin^{-1}(\frac{|u \times v|}{|u||v|}) \approx 1.3057 \ radians \approx 74.81°$

4. Lines

The symmetric equations of a line L through the point (x_0, y_0, z_0) and parallel to the direction vector $< a, b, c >$ are

$$\frac{x - x_0}{a} = \frac{y - y_0}{b} = \frac{z - z_0}{c}$$

The parametric equations of a line L through the point (x_0, y_0, z_0) and parallel to the direction vector $< a, b, c >$ are

$$\frac{x - x_0}{a} = \frac{y - y_0}{b} = \frac{z - z_0}{c} = t$$

Or

$$\begin{cases} x = x_0 + at \\ y = y_0 + bt \\ z = z_0 + ct \end{cases}$$

Example Let the line L through the point (1,1,-1) and parallel to the direction vector $< 1,2,4 >$.

Its parametric equations are

$$\begin{cases} x = 1 + t \\ y = 1 + 2t \\ z = -1 + 4t \end{cases}$$

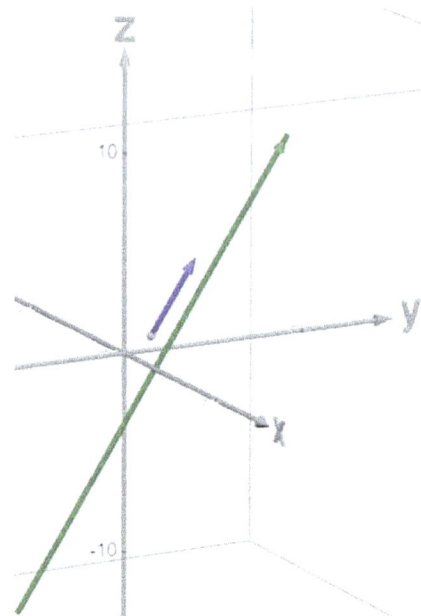

5. Planes

A linear equation in x, y, and z of the plane through the point (x_0, y_0, z_0) with the normal vector $n = <a, b, c>$ is

$$ax + by + cz + d = 0$$

Where $d = -(ax_0 + by_0 + cz_0)$

A scalar equation of the plane through the point (x_0, y_0, z_0) with normal vector

$n = <a, b, c>$ is

$$a(x - x_0) + b(y - y_0) + c(z - z_0) = 0$$

Example a plane through

the point (2,1,-2) with the normal vector $< 3,1,3 >$.

The plane's scalar equation is

$$3(x - 2) + (y - 1) + 3(z + 2) = 0$$

Or

$$3x + y + 3z - 1 = 0$$

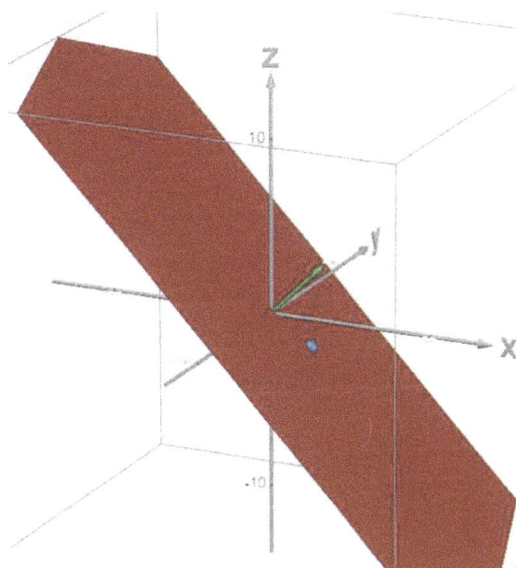

Calculate the distance between the point (5,2,8) and the plane

$$3x + y + 3z - 1 = 0$$

$$D = \frac{|3 \cdot 5 + 1 \cdot 2 + 3 \cdot 8|}{\sqrt{3^2 + 1^2 + 3^2}} = \frac{41}{\sqrt{19}} \approx 9.41$$

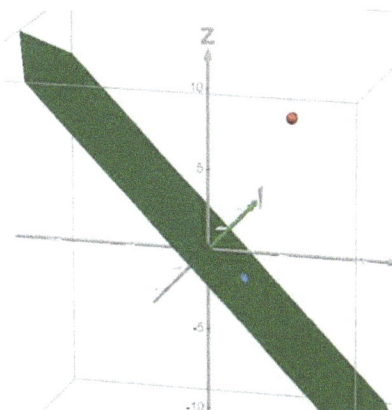

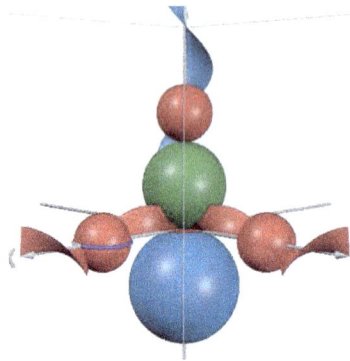

Appendix

Topics in College Algebra, Precalculus, Math Business and Economics I, Fundamental Math I, Fundamental Math II, and Elementary Statistics

Appendix

College Algebra – Math1314

1. Solve Equations
    a. Linear Equations
    b. Polynomial Equations
    c. Rational Equations
2. Function Operations
    a. Algebra of Functions
    b. Composition of Functions
    c. Transformations
3. Polynomial Functions and Their Graphs
    a. Understand the Domain and Range
    b. Evaluate a Polynomial Function
    c. Factoring
    d. Find the x-intercepts and y-intercepts
    e. Complex Numbers
    f. Division Theorems, Intermediate Value Theorem, Theorem about zeros of a Polynomial Function, Fundamental Theorem of Algebra
4. Rational Functions and Their Graphs
    a. Understand the Domain and Range
    b. Evaluate a Rational Function
    c. Factoring
    d. Find the x-intercepts and y-intercepts
    e. Horizontal and Vertical Asymptotes
5. Solve Inequations
    a. Linear Inequalities
    b. Polynomial Inequalities
    c. Rational Inequalities
6. Systems of Equations and Matrices
    a. System Equations in Two, Three, or More Variables
    b. System Linear Equations in Two, Three, or More Variables
    c. Algebra Methods (Substitution, Elimination) to Solve Systems of Linear Equations
    d. System Linear Equations in Matrices Representations
        i. Augment Matrices
        ii. Matrix Equations
7. Other
    a. Systems of Inequalities
    b. Linear Programming
    c. Partial Fractions

Precalculus Math2312/2412

1. Exponential and Logarithmic Functions

  a. Inverse Functions

  b. Domain and Range of Functions

  c. Exponential Functions

  d. Properties of Exponents

  e. Logarithmic Functions

  f. Properties of Logarithms

  g. Solve Exponential and Logarithmic Equations

2. Conic Sections

  a. Circle Equations

  b. Parabola Equations

  c. Ellipse Equations

  d. Hyperbola Equations

3. Trigonometric Functions

  a. Angles, Arcs, and Their Measures

  b. Domain and Range of Functions

  c. Sine, Cosine, Tangent, Cotangent, Secant, and Cosecant Functions

  d. Inverse Functions: Arcsine, Arccosine, Arctangent, Arccotangent, Arcsecant, Arccosecant

  e. Identities

    i. Fundamental: Reciprocal, Pythagorean, Quotient, Even-Odd Identities

    ii. From one Function to multiple terms: the argument of Sum, Difference, Double, and Half-Number

    iii. Product-to-Sum Identities

    iv. Sum-to-Product Identities

  f. Solve Trigonometric Equations

  g. Solve Triangles: Law of Sines and Cosines

4. Vectors

  a. Operations with Vectors

  b. Dot Product and Angle between Vectors

5. Other

  a. Parametric Equations and Graphs

  b. Polar Forms of Complex Numbers

  c. Powers and Roots of Complex Numbers

  d. Polar Equations and Graphs

  e. Limits

Math Business and Economic I – Math1324

1. Review Functions and Graphs

  a. Six Elementary Functions and Transformations

  b. Polynomial, Rational, Exponential, and Logarithmic Functions

  c. Regressions

2. Mathematics of Finance

  a. Simple Interest Formula

  b. Compound and Continuous Compound Interest Formulas

  c. Future Value of an Annuity Formula

  d. Present Value of an Annuity Formula

3. Systems of Linear Equations

     a. Augmented Matrices and Gauss-Jordan Elimination
     b. Matrix Equations
     c. Leontief Input-Output Analysis
4. Linear Inequalities and Linear Programming
     a. Systems of Linear Inequalities
     b. Objective Functions and Constraints
     c. Fundamental Theorem of Linear Programming
     d. Simplex Method
5. Logic, Sets, and Counting
     a. Logic
          i. Propositions
          ii. Compound Propositions: Negations, Disjunction, Conjunction, Conditional, and Biconditional
          iii. Logical Implication and Logical Equivalences
     b. Sets
          i. Elements/Members, the Empty/Null Set, Finite Sets, Infinite Sets, Subsets, Disjoint Sets, and Universal Set
          ii. Set Operations to form new sets: union, intersection, and complement
          iii. Display: Venn Diagrams
     c. Counting Principles
          i. Addition and Multiplication
          ii. Permutation
          iii. Combinations
6. Probability
     a. Sample Spaces, Events, and Probability
     b. Simple and Compound Events
     c. Conditional Probability and Bayes' Formula
     d. Random Variable, Probability Distribution, and Expected Value

Fundamental Math I – Math1350

1. Mathematical Problem-Solving Strategies
2. Logic and Sets
3. Numeration Systems, Whole Numbers, and Their Operations
4. Number Theory
5. Integers
6. Rational Numbers
7. Decimals
8. Algebraic Thinking

Fundamental Math II – Math1351

1. Probability
2. Data Analysis/Statistics
3. Euclidean Geometry
4. Congruence and Similarity
5. Measure, Area, and Volume
6. Transformations

Elementary Statistics – Math1342

1. Introduction to Statistics
2. Graphical Descriptions of Data
3. Numerical Descriptions of Data
4. Probability
5. Discrete Probability Distribution
6. Normal Probability Distribution
7. The Central Limit Theorem
8. Confidence Intervals for One and Two Samples
9. Hypothesis Testing for One, Two or More Populations

Technologies

1. Desmos Graphing Calculator
2. Desmos 3D Calculator
3. Geogebra Classic Calculator
4. Microsoft Word
5. Microsoft Excel
6. Adobe Acrobat
7. ChatGPT

Readings or References

1. Continued fraction - Wikipedia
2. A001203 - OEIS The online encyclopedia of integer sequences-Simple continued fraction expansion of Pi.
3. A002852 - OEIS Continued fraction for Euler's constant gamma.
4. Textbook: Algebra and Trigonometry, 5th Edition by Judith A Beecher, Judith A. Penna, and Marvin L. Bittinger, Pearson Education.
5. Textbook: A Problem-Solving Approach to Mathematics for Elementary School Teachers, 13th Edition by Billstein, Rick |Barbara Boschmans | Libeskind, Shlomo | Lott, Johnny; Publisher: Pearson Math & Statistics.
6. Textbook: College Mathematics for Business, Economics, Life Sciences, and Social Sciences 14th edition, by Barnett/Ziegler/Byleen; Publisher: Pearson
7. Textbook: Beginning Statistics, 3rd Edition by Carolyn Warren Wiley, Kimberly Denley, and Emily Atchley; Publisher: HAWKES LEARNING Math & Statistics
8. Textbook: A Graphical Approach to Precalculus with Limits, 7th Edition by John Hornsby, Margaret L. Lial, & Gary K. Rockswold, Publishing: Pearson Education
9. Textbook: Precalculus Mathematics: A Functional Approach, Fourth Edition by Karl J. Smith, Publisher: Brooks/Cole
10. Textbook: Calculus: Early Transcendentals 9th edition, by James Stewart, Daniel K. Clegg, Saleem Watson, Publisher: CENGAGE